# ADVANCES IN GEOTHERMAL ENERGY

Edited by **Basel I. Ismail**

## Advances in Geothermal Energy

http://dx.doi.org/10.5772/60623
Edited by Basel I. Ismail

### Contributors

Essam Aboud, Mike Middleton, Ioan Sarbu, Calin Sebarchievici, Rosa María Barragán, Víctor Manuel Arellano, David Nieva, Supri Soengkono, Ram Avtar

**CBS,** Edition **2017**

### Published by InTech

Janeza Trdine 9, 51000 Rijeka, Croatia



**Publishing Process Manager** Sandra Bakic

**Technical Editor** InTech DTP Team

**Cover Designer** InTech Design Team

Additional hard copies can be obtained from orders@intechopen.com

Advances in Geothermal Energy, Edited by Basel I. Ismail
p. cm.
ISBN 978-953-51-2241-8

INTECH

open science | open minds

# Contents

# Preface

Geothermal energy means the natural heat energy from the Earth. The source of geothermal energy is the continuous heat energy flux flowing from the interior of the Earth towards its surface. The geothermal resources of the Earth are enormous; for example, the part of geothermal energy stored at a depth of 3 km is estimated to be 1,194,444,444 TWh which is much larger compared to all fossil fuel resources combined, whose energy equivalent is estimated to be 1, 010,361 TWh. Geothermal energy resources vary geographically from one location to another, depending on the depth and temperature of the resource, the rock chemical composition and the abundance of ground water. Unlike other conventional and renewable energy sources, geothermal energy has unique features; namely, it is available, stable at all times throughout the year, independent of weather conditions, and has an inherent storage capability. Geothermal energy is also considered to be an environmentally friendly clean energy source that could significantly contribute to the reduction of GHG emissions when utilized for electrical power generation. It was estimated that the world net electricity demand is going to increase by approximately 85% from 2004 to 2030, rising from 16,424 TWh (in 2004) to 30,364 TWh in the year 2030 so that the utilization of geothermal energy for power generation continues to be an attractive solution especially with the new discoveries of innovative technological methods of drilling and power generation cycles. The utilization of geothermal energy can also be used for direct heating applications.

Due to its important utilization and future prospects, various interesting topics of research related to geothermal energy are covered in this book. This book is the result of contributions from several researchers and experts worldwide. It is hoped that the book will become a useful source of information and basis for extended research for researchers, academics, policy makers, and practitioners in the area of geothermal energy.

This book contains six chapters. Chapter one presents a detailed theoretical study, economic analysis (using different indicators), numerical simulations, and experimental investigations of ground-source heat pump (GSHP) systems. The main performance parameters (energy efficiency and $CO_2$ emissions) of radiator and radiant floor heating systems connected to a ground-coupled heat pump are compared. Moreover in this chapter, two numerical simulation models of the useful thermal energy and the system coefficient of performance in heating mode are developed using the TRNSYS software. Finally, the simulations obtained from TRNSYS program are analyzed and compared to experimental measurements.

Chapter two primarily discusses various challenges and opportunities in geothermal energy policies of Indonesia and Philippines in order to adopt them to the Japanese society needs in the future. Also, a review of the processes utilized for policy implementation is presented, looking at the effectiveness of certain policy instruments. Community based development of direct uses of geothermal energy, an area that has not been analyzed adequately in the past, was also assessed in this chapter.

Chapter three reviews geothermal heat generation in crystalline rocks and possible influences on overlying sedimentary basins in Western Australia. This chapter also outlines the regions containing higher than normal levels of uranium, thorium and potassium adjacent to the sedimentary basins, and propose correlations between these regions to elevated heat flow in the sedimentary basins.

In chapter four, a methodology based on the variations of geochemical and production data of wells overtime, was described. This methodology has proved to be successful to investigate the response of geothermal reservoirs to exploitation and its use was illustrated in some examples for Mexican geothermal fields. The results from this approach together with results from other disciplines provide support in field management on delineating optimal exploitation strategies to prolong the geothermal reservoir lifetime in a sustainable way.

Chapter five discusses simple approach to use airborne magnetic data for the investigation of high-temperature geothermal resources in volcanic setting. The physical background of airborne magnetic survey is discussed in a way that is simple and easy to understand. Examples are given for interpretations of real airborne magnetic data observed at two different magnetic latitudes, the North Island of New Zealand and the Java Island of Indonesia. This chapter is aimed to provide researchers with sufficient degree of confidence in organising and/or running investigation of high-temperature geothermal reservoirs using airborne magnetic data.

Finally, chapter six presents detailed information and aspects with examples related to different surveys methods of geothermal reservoirs. The presented information is resourceful for researchers and practitioners in the area of explorations of geothermal reservoirs for commercially viable power generation systems.

I would like to thank all chapter authors for their efforts and the quality of the chapters presented. Also, I would like to thank Ms. Sandra Bakic from InTech publisher for her excellent efforts in managing the publication process of this book.

**Dr. Basel I. Ismail, P.Eng.**
Associate Professors and Chair
Department of Mechanical Engineering
Faculty of Engineering
Lakehead University
Thunder Bay, Ontario, Canada

Chapter 1

# Using Ground-Source Heat Pump Systems for Heating/ Cooling of Buildings

Ioan Sarbu and Calin Sebarchievici

Additional information is available at the end of the chapter

http://dx.doi.org/10.5772/61372

**Abstract**

This chapter mainly presents a detailed theoretical study and experimental investigations of ground-source heat pump (GSHP) technology, concentrating on the ground-coupled heat pump (GCHP) systems. A general introduction on the GSHPs and its development, and a description of the surface water (SWHP), ground-water (GWHP), and ground-coupled heat pumps are briefly performed. The most typical simulation and ground thermal response test models for the vertical ground heat exchangers (GHEs) currently available are summarized. Also, a new GWHP using a heat exchanger with special construction, tested in laboratory, is well presented. The second objective of the chapter is to compare the main performance parameters (energy efficiency and $CO_2$ emissions) of radiator and radiant floor heating systems connected to a GCHP. These performances were obtained with site measurements in an office room. Furthermore, the thermal comfort for these systems is compared using the ASHRAE Thermal Comfort program. Additionally, two numerical simulation models of useful thermal energy and the system coefficient of performance ($COP_{sys}$) in heating mode are developed using the TRNSYS (Transient Systems Simulation) software. Finally, the simulations obtained in TRNSYS program are analysed and compared to experimental measurements.

**Keywords:** Geothermal energy, heat pump, ground heat exchanger, energy efficiency, radiator heating, radiant floor heating, experimental measurements, system performance, simulation models

## 1. Introduction

An economical strategy of a sustainable development imposes certainly to promote efficiency and a rational energy use in buildings as the major energy consumer in Romania and the other member states of the European Union (EU). Energy consumption patterns EU reveal that buildings are the greatest energy consumer, consuming 41% of energy, followed by industry and transportation consuming approximately 30% [1]. Buildings represent the biggest and most cost-effective potential for energy savings. Also, studies have shown that saving energy is the most cost-effective method to reduce greenhouse gas (GHG) emissions. At present, heat use is responsible for almost 80% of the energy demand in houses and utility buildings for space heating and hot-water generation, whereas the energy demand for cooling is growing year after year.

In order to realise the ambitious goals for the reduction of fossil primary energy consumption and the related $CO_2$ emissions to reach the targets of the Kyoto–protocol besides improved energy efficiency, the use of renewable energy in the existing building stock have to be addressed in the near future [2].

On 23 April 2009, the European Parliament and the Council adopted the Renewable Energy Directive 2009/28/EC. It establishes a common framework for the promotion of energy from renewable sources. This directive opens up a major opportunity for further use of heat pumps for heating and cooling of new and existing buildings. Heat pumps enable the use of ambient heat at useful temperature level need electricity or other energy form to function [2]. Furthermore, EU member states must stimulate the transformation of existing buildings undergoing renovation into nearly zero-energy buildings (nZEBs). Conversion to heating and cooling systems based on ground-source heat pumps and air-to-water heat pumps is a well-proven measure to approach nZEB requirements.

Ground-source heat pump (GSHP) systems use the ground as a heat source/sink to provide space heating and cooling as well as domestic hot-water. The GSHP technology can offer higher energy efficiency for air-conditioning compared to conventional air-conditioning (A/C) systems because the underground environment provides higher temperature for heating and lower temperature for cooling and experiences less temperature fluctuation than ambient air temperature change [3]. To date, the GSHP systems have been widely used in both residential and commercial buildings. It is estimated that the GSHP system installations have grown continuously on a global basis with the range from 10 to 30% annually in recent years [4].

A ground-coupled heat pump (GCHP) system consists of a conventional heat pump coupled with a ground heat exchanger (GHE) where water or a water-antifreeze solution exchanges heat with the ground. The GHE may be a simple pipe system buried in the ground; it may also comprise a horizontal collector or, more commonly, borehole heat exchanger (BHE) drilled to a depth between 20 and 300 m with a diameter of 100–200 mm [5].

The widespread distribution of heat pumps as single generators in heating systems has mainly been in new, rather isolated buildings, thus having limited unit loads. This has enabled the use of low-temperature terminal units, such as fan coil units and, especially, radiant systems [6]. After the introduction of plastic piping water-based radiant heating and cooling with pipes embedded in room surfaces (floor, wall, and ceiling), the application increased significantly worldwide. Due to the large surfaces needed for heat transfer, the systems work with low water temperature for heating and high water temperature for cooling. However, in order to extend the use of these types of generators and benefit from their energy efficiency to reach

http://dx.doi.org/10.5772/61372

the targets of 20-20-20, it is also compulsory to work with radiators, which were the most commonly used terminal units in heating systems in the past.

This chapter mainly presents a detailed theoretical study and experimental investigations of GSHP technology, concentrating on the GCHP systems. Initially, the operation principles of a heat pump are described and their energy, economic and environmental performances are defined, showing the opportunity to implement the heat pump in a heating/cooling system. Then, a general introduction on the GSHPs and its development, and a description of the surface water (SWHP), ground-water (GWHP), and ground-coupled (GCHP) heat pumps are briefly performed. The most typical simulation and ground thermal response test models for the vertical GHEs currently available are summarized, including the heat transfer processes outside and inside the boreholes. Additionally, a new GWHP using a heat exchanger with special construction, tested in laboratory, is well presented. The second objective of the chapter is to compare the main performance parameters (energy efficiency and $CO_2$ emissions) of radiator and radiant floor heating systems connected to a GCHP. These performances were obtained with site measurements in an office room. Furthermore, the thermal comfort for these systems is compared using the ASHRAE Thermal Comfort program. Additionally, two numerical simulation models of useful thermal energy and the system coefficient of performance ($COP_{sys}$) in heating mode are developed using the TRNSYS (Transient Systems Simulation) software. Finally, the simulations obtained in TRNSYS program are analysed and compared to experimental measurements.

## 2. Operation Principle of a Heat Pump

A heat pump (HP) is a thermal installation that is based on a reverse Carnot thermodynamic cycle (consumes drive energy and produces a thermal effect). Any HP moves (pumps) heat $E_S$ from a source with low temperature $t_s$ to a source with a high temperature $t_u$, consuming the drive energy $E_D$. A heat source can be:

- a gas or air (outdoor air, warm air from ventilation, hot gases from industrial processes);
- a liquid called generic water: surface water (river, lake, or sea), ground-water, or discharged hot-water (domestic, technologic, or recirculated in cooling towers); or
- ground, with the advantage of accessibility.
- *Heat consumer*. The heat pump yields thermal energy at a higher temperature, depending on the application of the heat consumer. This energy can be used for:
- space heating, which is related to low temperature heating systems: radiant panels (floor, wall, ceiling, or floor-ceiling), warm air, or convective systems; or
- water heating (pools, domestic or technologic hot-water);

The heat consumer is recommended to be associated with a cold consumer. This can be performed with either a reversible (heating–cooling) or a double effect system. In cooling mode, a heat pump operates exactly like central air-conditioning.

- *Drive energy*. Heat pumps can be used to drive different energy forms:
- electrical energy (electro-compressor);
- mechanical energy (mechanical compression with expansion turbines);
- thermo-mechanical energy (steam ejector system);
- thermal energy (absorption cycle); or
- thermo-electrical energy (Péltier effect).

The GSHPs are those with electro-compressor. The process of elevating low temperature heat to over 38°C and transferring it indoors involves a cycle of evaporation, compression, condensation, and expansion (Figure 1). A non-CFC refrigerant is used as the heat-transfer medium, which circulates within the heat pump [7].

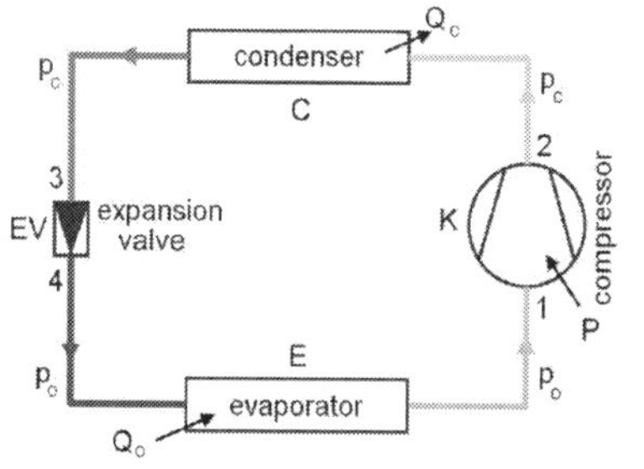


**Figure 1.** Schematic of single-stage compression refrigeration system

## 3. Performances and $CO_2$ emissions of a heat pump

The opportunity to implement a HP in a heating/cooling system results on the basis of energy indicators and economic analysis.

### 3.1. Energy efficiency

#### *3.1.1. Coefficient of performance*

The operation of a heat pump is characterised by the coefficient of performance (COP) defined as the ratio between useful thermal energy $E_t$ and electrical energy consumption $E_{el}$:

$$COP = \frac{E_t}{E_{el}} \quad (1)$$

http://dx.doi.org/10.5772/61372

If both usable energy and consumed energy are summed during a season (year) is obtained by Eq. (1) seasonal coefficient of performance ($COP_{seasonal}$) or average COP over a heating (cooling) season, which is often indicated as seasonal performance factor (SPF) or annual efficiency.

In the heating operate mode, the heat pump COP is defined by equation:

$$\mathrm{COP}_{hp} = \frac{Q_{HP}}{P_e} \tag{2}$$

where $Q_{HP}$ is the thermal power (capacity) of heat pump, in W; $P_e$ is the electric power consumed by the compressor of heat pump, in W.

In the cooling mode, a HP operates exactly like a central air conditioner. The energy efficiency ratio (EER) is analogous to the COP but tells the cooling performance. The $EER_{hp}$, in Btu/(Wh), is defined as:

$$\mathrm{EER}_{hp} = \frac{Q_0}{P_e} \tag{3}$$

where $Q_0$ is the cooling power of heat pump, in British Thermal Unit per hour (Btu/h); $P_e$ is the compressor power, in W.

The coefficient of performance of heat pump in cooling mode is obtained by the following equation:

$$\mathrm{COP}_{hp} = \frac{\mathrm{EER}_{hp}}{3.412} \tag{4}$$

where value 3.412 is the transformation factor from Watt in Btu/h.

Figure 2 illustrates the COP variation of heat pumps in the heating operation mode, according to the source temperature $t_s$ and the temperature at the consumer $t_u$ [5].

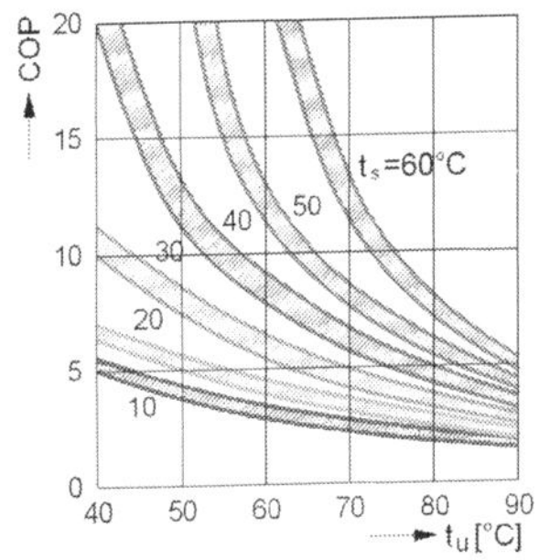


**Figure 2.** Efficiency variation of heat pumps

The GSHP systems intended for ground-water or oven-system applications have heating COP ratings ranging from 3.0 to 4.0 and cooling EER ratings between 11.0 and 17.0. Those systems intended for closed-loop applications have COP ratings between 2.5 and 4.0 and EER ratings ranging from 10.5 to 20.0 [8]. The characteristic values of the SPF of modern GSHPs are commonly assumed to be approximately 4, meaning that four units of heat are gained per unit of consumed electricity.

The sizing factor (SF) of the HP is defined as the ratio of the heat pump capacity $Q_{HP}$ to the maximum heating demand $Q_{max}$:

$$\mathrm{SF} = \frac{Q_{HP}}{Q_{\max}} \tag{5}$$

The SF can be optimized in terms of energy and economics, depending on the source temperature and the used adjustment schedule.

*3.1.2. Profitability and capabilities of heat pump*

The factors that can affect the life-cycle efficiency of a HP are (1) the local method of electricity generation; (2) the local climate; (3) the type of heat pump (ground or air source); (4) the refrigerant used; (5) the size of the heat pump; (6) the thermostat controls; and (7) the quality of work during installation.

Considering that the HP has over-unit efficiency, to evaluate the consumed primary energy uses a synthetic indicator [5]:

$$\eta_s = \eta_g \mathrm{COP}_{hp} \tag{6}$$

in which:

$$\eta_g = \eta_p \eta_t \eta_{em} \tag{7}$$

where $\eta_g$ is the global efficiency and $\eta_p$, $\eta_t$ and $\eta_{em}$ are the electricity production, the transportation and the electromotor efficiency, respectively.

For justifying the use of a heat pump, the synthetic indicator has to satisfy the condition $\eta_s > 1$. Additionally, the use of a heat pump can only be considered if the $\mathrm{COP_{hp}} > 2.78$.

The COP of a heat pump is restricted by the second law of thermodynamics:

- in heating mode:

$$\mathrm{COP} \le \frac{t_u}{t_u - t_s} = \varepsilon_C \tag{8}$$

- in the cooling mode:

$$\mathrm{COP} \le \frac{t_s}{t_u - t_s} \tag{9}$$

where $t_u$ and $t_s$ are the absolute temperatures of the hot environment (condensation temperature) and the cold source (evaporation temperature), respectively, in K.

The maximum value $\varepsilon_C$ of the efficiency can be obtained in the reverse Carnot cycle.

### 3.2. Economic indicators

In the economic analysis of a system, different methods could be used to evaluate the systems. Some of them are: the present value (PV) method, the net present cost (NPC), the future value (FV) method, the total annual cost (TAC) method, the total updated cost (TUC) method, the annual life cycle cost (ALCC), and other methods.

- The PV of a future payment can be calculated using the equation [9]:

$$PV = \frac{C}{(1+i)^{\tau}} \tag{10}$$

where $C$ is the payment/cost on a given future date; $\tau$ is the number of periods to that future date; $i$ is the discount (interest) rate. Therefore, PV is the present value of a future payment that occurs at the end of the $\tau$-th period.

Similarly, the PV of a stream of costs with a specified number of fixed periodic payments can be expressed as:

$$PV = u_r C = \frac{C}{CRF} \tag{11}$$

where the update rate $u_r$ is defined as:

$$u_r = \sum_{n=1}^{\tau} \frac{1}{(1+i)^n} = \frac{(1+i)^{\tau} - 1}{i(1+i)^{\tau}} = \frac{1}{CRF} \tag{12}$$

where $C$ is the periodic payment that occurs at the end of each period; $n$ is the number of periods (years); $CRF$ is the capital recovery factor.

- Another economic indicator is total updated cost:

$$TUC = I_0 + \sum_{n=1}^{\tau} \frac{C}{(1+i)^n} \tag{13}$$

where $I_0$ is the initial investment cost, in the operation beginning date of the system; $C$ is annual operation and maintenance cost of the system; $i$ is the discount (inflation) rate; $\tau$ is the number of years for which is made update (20 years).

Taking into account Eq. (12), Eq. (13) gets the form:

$$TUC = I_0 + u_r C \tag{14}$$

- Usually, the HP achieves a fuel economy $\Delta C$ (operating costs) comparatively of the classical system with thermal station (TS), which is dependent on the type of HP. On the other hand, the HP involve an additional investment $I_{HP}$ from the classical system $I_{TS}$, which produces the same amount of heat [2].

Thus, it can be determined the recovery time $RT$, in years, to increase investment, $\Delta I = I_{HP} - I_{TS}$, taking into account the operation saving achieved through low fuel consumption $\Delta C = C_{TS} - C_{HP}$:

$$RT = \frac{\Delta I}{\Delta C} \leq RT_n \tag{15}$$

where $RT_n$ is normal recovery time.

It is estimated that for $RT_n$ a number 8–10 years is acceptable, but this limit varies depending on the country's energy policy and environmental requirements.

### 3.3. Calculation of greenhouse gas emissions

Due to the diversity in each country with respect to heating practices, direct geothermal energy use by GSHPs, and primary energy sources for electricity, country-specific calculations are provided.

The annual heating energy provided by GSHPs is defined as $E_t$. The annual primary energy consumption from heat pump electricity use is then:

$$E_{el} = \frac{E_t}{\text{SPF}} \tag{16}$$

Because heat pump electricity consumption is considered the most important source for greenhouse gas (GHGs) emissions [10], other potential contributors (e.g., heat pump life cycle, heat pump refrigerant, and borehole construction) are neglected. Applying an emission factor $g_p$, in kg $CO_2$/kWh, the annual GHG emissions $C_{GSHP}$, in kg $CO_2$, from GSHP operation can be obtained:

$$C_{\text{GSHP}} = g_p E_{el} \tag{17}$$

The emission factor typically varies among different countries and characterises the GHG intensity of electricity production. Note that although carbon dioxide ($CO_2$) represents the most important greenhouse gas, there exist several other compounds that contribute similarly to climate change. Their combined impact is commonly normalised to the specific effect of $CO_2$, and all emissions are expressed in $CO_2$ equivalents. For the sake of readability, however, the emissions are expressed only in kg $CO_2$.

Thus, the $CO_2$ emissions $C_{CO_2}$ of the GSHP during its operation can be evaluated with the following equation:

$$C_{CO_2} = g_{el} E_{el} \tag{18}$$

where $g_{el}$ is the specific $CO_2$ emission factor for electricity. The average European $CO_2$ emission factor for electricity production is 0.486 kg $CO_2$/kWh and for Romania is 0.547 kg $CO_2$/kWh [11].

# 4. Ground-Source heat pump systems

## 4.1. Generalities

Heat pumps are classified by (1) the heat source and sink; (2) the heating and cooling distribution fluids; and (3) the thermodynamic cycle. The following classifications can be made according to:

- function: heating, cooling, domestic hot-water (DHW) heating, ventilation, drying, heat recovery, etc.
- heat source: ground, ground-water, water, air, exhaust air etc.
- heat source (intermediate fluid)-heat distribution: air-to-air, air-to-water, water-to-water, antifreeze (brine)-to-water, direct expansion-to-water, etc.

Recently, the GSHP system has attracted more and more attention due to its superiority of high energy efficiency and environmental friendliness [4,5,12]. Renewable forms of energy such as solar, wind, biomass, hydro, and earth energy produce low or no GHG emissions. The temperature of the ground is fairly constant below the frost line. The ground is warmer in the middle of winter and cooler in the middle of summer than the outdoor air. Thus, the ground is an efficient heat source. A GSHP system includes three principle components: (1) a ground connection subsystem, (2) heat pump subsystem, and (3) heat distribution subsystem.

The GSHPs comprise a wide variety of systems that may use ground-water, ground, or surface water as heat sources or sinks. These systems have been basically grouped into three categories by ASHRAE [13]: (1) ground-water heat pump (GWHP) systems, (2) surface water heat pump (SWHP) systems, and (3) ground-coupled heat pump (GCHP) systems. The schematics of these different systems are shown in Figure 3. Many parallel terms exist: geothermal heat pump (GHP), earth energy system (EES), and ground-source system (GSS).

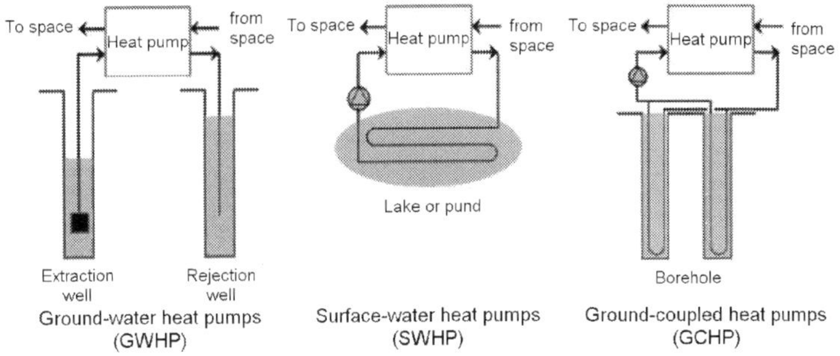


**Figure 3.** Schematic of different ground-source heat pumps

Among the various GSHP systems, the vertical GCHP system has attracted the greatest interest in research field and practical engineering. Several literature reviews on the GCHP technology have been reported [14].

In a GCHP system, heat is extracted from or rejected to the ground via a closed-loop, i.e., ground heat exchanger (GHE), through which pure water or antifreeze fluid circulates. The GHEs commonly used in the GCHP systems typically consist of HDPE pipes which are installed in either vertical boreholes (called vertical GHE) or horizontal trenches (horizontal GHE) [3]. In direct expansion systems, the heat stored in the ground is absorbed directly by the working fluid (refrigerant). This results in an increased coefficient of performance. Horizontal GHEs are mainly used with this system.

### 4.2. Description of SWHP systems

Surface water bodies can be very good heat source and sinks, if properly used. The maximum density of water occurs at 4.0°C, not at the freezing point of 0°C. This phenomenon, in combination with the normal modes of heat transfer to and from takes, produces temperature profile advantageous to efficient heat pump operation. In some cases, lakes can be the very best water supply for cooling. Various water circulation systems are possible and several of the more common are presented [13].

The closed-loop systems consist of water-to-air or water-to-water heat pumps connected to a piping network placed in a lake, river, or other open body of water. A pump circulates water or a water/antifreeze solution through the heat pump water-to-refrigerant heat exchanger and the submerged piping loop, which transfers heat to or from the body of water.

Open-loop systems can use surface water bodies the way cooling towers are used, but without the need for fan energy or frequent maintenance. In warm climates, lakes can also serve as heat sources during winter heating mode, but in colder climates where water temperatures drop below 7°C, closed-loop systems are the only viable option for heating.

Lake water can be pumped directly to water-to-air or water-to-water heat pumps or through an intermediate heat exchanger that is connected to the units with a closed piping loop. Direct systems tend to be smaller, having only a few heat pumps. In deep lakes (12 m or more), there is often enough thermal stratification throughout the year that direct cooling or precooling is possible. Water can be pumped from the bottom of deep lakes through a coil in the return air duct. Total cooling is possible if water is 10°C or below. Precooling is possible with warmer water, which can then be circulated through the heat pump units.

Advantages of closed-loop SWHPs are (1) relatively low cost because of reduced excavation costs, (2) low pumping energy requirements, and (3) low operating cost. Disadvantages are (1) the possibility of coil damage in public lakes and (2) wide variation in water temperature with outdoor conditions.

### 4.3. Description of GWHP systems

A GWHP system removes ground-water from a well and delivers it to a heat pump (or an intermediate heat exchanger) to serve as a heat source or sink [13]. One widely used design places a central water-to-water heat exchanger between the ground-water and a closed water loop, which is connected to water-to-air heat pumps in the building. A second possibility is to circulate ground-water through a heat recovery chiller, and to heat and cool the building with a distributed hydronic loop.

Direct systems (in which ground-water is pumped directly to the heat pump without an intermediate heat exchanger) are not recommended except on the very smallest installations. Although some installations of this system have been successful, others have had serious difficulty even with ground-water of apparently benign chemistry. The specific components for handling ground-water are similar. The primary items include (1) wells (supply and, if required, injection), (2) a well pump (usually submerged), and (3) a ground-water heat exchanger. The use of a submerged pump avoids the possibility of introducing air or oxygen into the system. A back-washable filter should also be installed. The injection well should be located from 10 to 15 m in the downstream direction of the ground-water flow.

In an open-loop system, the intermediate heat exchanger between the refrigerant and the ground-water is subject to fouling, corrosion, and blockage. The required flow rate through the intermediate heat exchanger is typically between 0.027 and 0.054 l/s. The ground-water must either be reinjected into the ground by separate wells or discharged to a surface system such as a river or lake. The drill diameter should be at least 220 mm (larger for sandy conditions to prevent sand entry).

The ground-water flow rate $G$ must be capable of delivering the full capacity required from the heat source. This depends on the evaporator cooling power $Q_0$ and the water cooling degree and is given by the following equation:

$$G = \frac{Q_0}{\varrho_w c_w (t_{wi} - t_{we})} \tag{19}$$

where $\varrho_w$ is the water density; $c_w$ is the specific heat of water; and $t_{wi}$ and $t_{we}$ are the water temperatures at the heat pump inlet and the heat pump outlet, respectively.

Table 1 summarises the calculated COP values of GWHP and SWHP systems, operating as water-to-water heat pumps.

| **Water temperature at evaporator inlet** *$t_s$ [°C]* | **Water temp. at condenser outlet, $t_u$ [°C]** | | | | |
|---|---|---|---|---|---|
| 30 | 35 | 40 | 45 | 50 | |
| 5 | 4.55 | 4.10 | 3.70 | 3.40 | 3.15 |
| 10 | 5.30 | 4.65 | 4.15 | 3.75 | 3.45 |
| 15 | 6.25 | 5.35 | 4.70 | 4.20 | 3.85 |
| 20 | 7.70 | 6.35 | 5.45 | 4.80 | 4.30 |
| 25 | 9.95 | 7.80 | 6.45 | 5.55 | 4.85 |
| 30 | 14.10 | 10.10 | 7.95 | 6.55 | 5.60 |

**Table 1.** The COP of water-to-water GWHP and SWHP systems

The "Geotherm" system [5] uses a specially built heat exchanger (Figure 4) placed in an extraction well with a 1.0 m diameter and a depth of 2.0 m. The heat exchanger is mounted between a GCHP and a ground-water source with a reduced flow rate and of any water quality. This heat exchanger consists of a set of four coaxial coils made of HDPE tubes with a diameter of 25 mm, immersed in a cylindrical reservoir made of glass fibre reinforced resins (0.8 m diameter and 1.2 m height) supplied with ground-water at the bottom side.

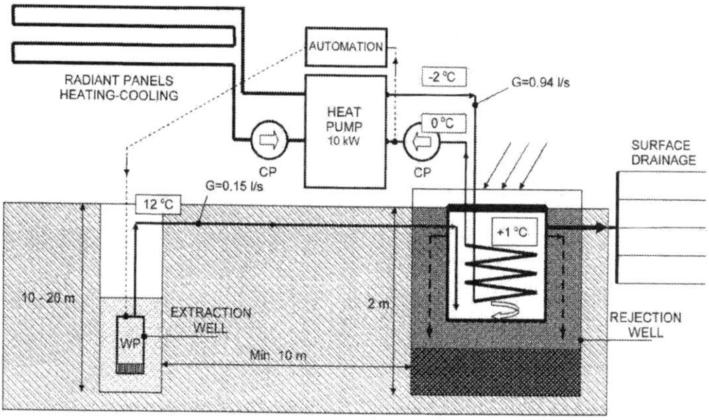


**Figure 4.** Schematic of "Geotherm" GWHP system

The heat pump used in conjunction with the intermediate heat exchanger is a GCHP system of 10 kW with a COP of 4.

The secondary circuit of the heat exchanger (towards the heat pump) circulates an antifreeze solution (glycol 20%), which enters the heat pump at 0°C and leaves at –2°C, transported by a circulation pump with a flow rate of 0.94 l/s. The glycol flow in the tubes is ensured by the circulation pump within the heat pump. Outside of the coils, the ground-water from the cylindrical reservoir is involved in a flow among the spires of coils by a submersible pump. The relatively small pressure loss on the secondary circuit of the heat exchanger allows the use of a reduced power circulation pump for the glycol [2].

In the primary circuit of the heat exchanger (outside the tubes), ground-water enters with a temperature of 12°C and is evacuated to approximately 1°C (in heating mode). Because the temperature drop is 11°C, compared to 4°C in the usual systems, it is possible to obtain the same thermal power with a ground-water flow rate nearly three times lower. The pressure loss on the primary circuit of the heat exchanger is 26 kPa for the mentioned flow rate. The heat exchange is realized mainly by the ground-water supply, and the heat exchanged directly with the ground around the extraction well is also important. The heat transfer surface is 20 $m^2$, and the heat transfer coefficient is 154 $W/m^2K$.

The ground-water is then evacuated through the top of the heat exchanger by gravity in the rejection well. If the rejection well cannot retrieve all of the ground-water flow rate, surface drainage through a network of perforated pipes buried at 50–80 cm or another evacuation solution (lake, river, or sewer) is recommended.

Regardless of the outdoor air and ground temperature, the heat pump will always operate at the same optimum temperatures because of the automation. The automation starts the ground-water inlet (electro-valve or submersible pump) only when the return water–glycol temperature goes below 1°C. The ground-water flow rate is limited to 4–12 l/min depending on the thermal power of the heat pump (4–12 kW).

During the summer, the intermediate heat exchanger can operate in a passive cooling mode in which the heat pump only produces domestic hot-water using heat recovered from the air-conditioned space. In this case, the heat carrier from the heaters is transported with the circulation pump directly to the intermediate heat exchanger.

### 4.4. Description of GCHP systems

The GCHP is a subset of the GSHP and is often called a closed-loop heat pump. A GCHP system consists of a reversible vapour compression cycle that is linked to a GHE buried in the soil (Figure 4). The GCHP is further subdivided according to GHE type: horizontal GHE and vertical GHE.

#### *4.4.1. Types of horizontal GHEs*

Horizontal GHEs (Figure 5) can be divided into at least three subgroups: single-pipe, multiple-pipe, and spiral. Single-pipe horizontal GHEs consist of a series of parallel pipe arrangements laid out in trenches. Typical installation depths in Europe vary from 0.8 to 1.5 m. Antifreeze fluid runs through the pipes in a closed system. The values of the specific extraction/rejection power $q_E$ for ground [15] are given in Table 2. For a specific power of extraction/rejection, $q_E$ can be obtained from required ground area [16]:

$$A = \frac{Q_0}{q_E} \tag{20}$$

where $Q_0 = Q_{HP} - P_e$ is the cooling power of heat pump.

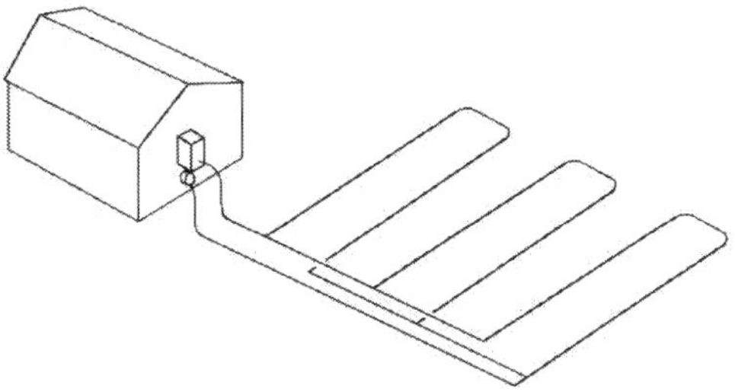

**Figure 5.** Horizontal ground heat exchanger

| No. | Type of ground | $q_E$ [W/m²] |
|---|---|---|
| 1 | Dry sandy | 10...15 |
| 2 | Moist sandy | 15...20 |
| 3 | Dry clay | 20...25 |
| 4 | Moist clay | 25...30 |
| 5 | Ground with ground-water | 30...35 |

**Table 2.** Specific extraction/rejection power for ground

To save required ground area, some special GHEs have been developed [7]. Multiple pipes (two, four, or six), placed in a single trench, can reduce the amount of required ground area. The spiral loop (Figure 6) is reported to further reduce the required ground area. This consists of pipe unrolled in circular loops in trenches with a horizontal configuration. For the horizontal spiral loop layout, the trenches are generally a depth of 0.9 to 1.8 m. The distance between coil tubes is of 0.6–1.2 m. The length of collector pipe is of 125 m per loop (up to 200 m). The ends of parallel coils 1 are arranged by a manifold-collector 2 in a heart 3, and then the antifreeze fluid is transported by main pipes 4 at heat pump. Disadvantages of the horizontal systems are: (1) these systems are more affected by ambient air temperature fluctuations because of their proximity to the ground surface, and (2) the installation of the horizontal systems needs much more ground area than vertical system.

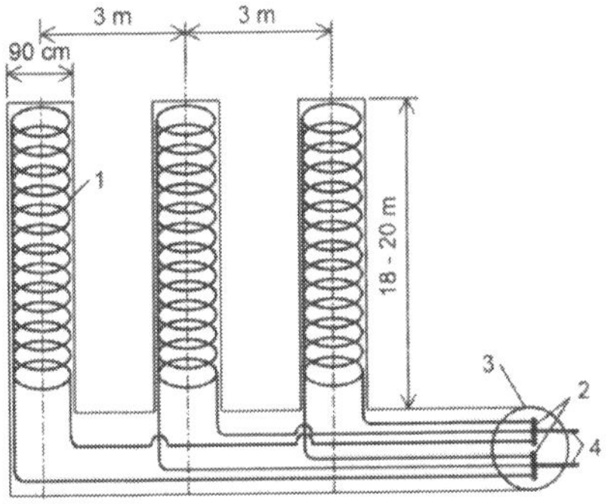


**Figure 6.** Spiral ground coil

### *4.4.2. Types of vertical GHEs*

There are two basic types of vertical GHEs or borehole heat exchangers (BHE): U-tube and concentric- (coaxial-) tube system configurations (Figure 7). BHEs are widely used when there is a need to install sufficient heat exchanger capacity under a confined surface area, such as when the earth is rocky close to the surface, or where minimum disruption of the landscape is desired. The U-tube vertical GHE may include one, tens, or even hundreds of boreholes, each containing single or double U-tubes through which heat exchange fluid are circulated. Typical U-tubes have a nominal diameter in the range of 20–40 mm and each borehole is normally 20–200 m deep with a diameter ranging from 100 to 200 mm [17]. Concentric pipes, either in a very simple method with two straight pipes of different diameters or in complex configurations, are commonly used in Europe. The borehole annulus is generally backfilled with some special material (grout) that can prevent contamination of ground-water.

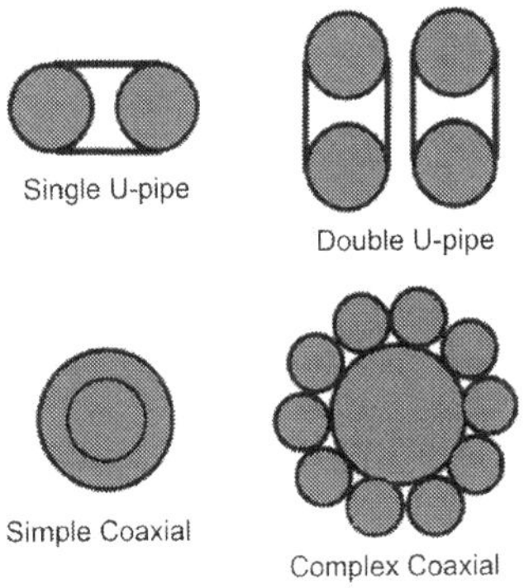


**Figure 7.** Common vertical GHE designs

A typical borehole with a single U-tube is illustrated in Figure 8. The required borehole length $L$ can be calculated by steady-state heat transfer equation as follows [13]:

$$L = \frac{qR_g}{t_g - t_f} \tag{21}$$

where $q$ is the heat transfer rate, in kW; $t_g$ is the ground temperature, in K; $t_f$ is the heat carrier fluid (antifreeze, refrigerant) temperature, in K; $R_g$ is the effective thermal resistance of ground per unit length, in (mK)/kW.

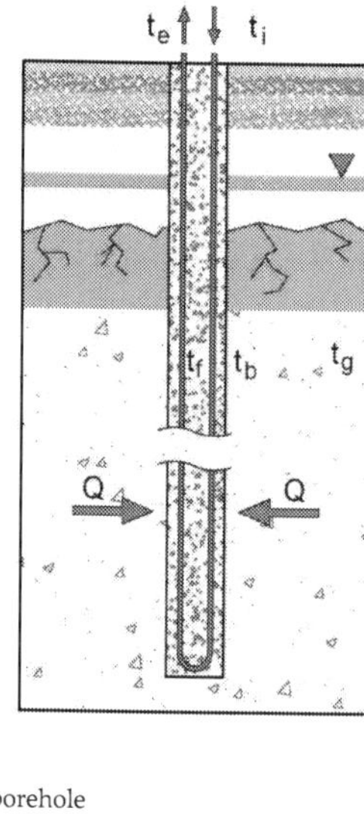


**Figure 8.** Schematic of a vertical grouted borehole

The GHE usually are designed for the worst conditions by considering that these needs to handle three consecutive thermal pulses of various magnitude and duration: yearly average ground load $q_a$ for 20 years, the highest monthly ground load $q_m$ for 1 month, and the peak hourly load $q_h$ for 6 h. The required borehole length to exchange heat at these conditions is given by [18]:

$$L = \frac{q_h R_b + q_a R_{20a} + q_m R_{1m} + q_h R_{6h}}{t_g - (t_f + \Delta t_g)} \tag{22}$$

where $R_b$ is the effective borehole thermal resistance; $R_{20a}$, $R_{1m}$, $R_{6h}$ are the effective ground thermal resistances for 20 years, 1 month, and 6 h thermal pulses; $\Delta t_g$ is the increase of temperature because of the long-term interference effect between the borehole and the adjacent boreholes. Alternative methods of computing the thermal borehole resistance are presented by Bernier [18] and Hellström [19].

Advantages of the vertical GCHP are that it (1) requires relatively small ground area, (2) is in contact with soil that varies very little in temperature and thermal properties, (3) requires the smallest amount of pipe and pumping energy, and (4) can yield the most efficient GCHP system performance. Disadvantage is the higher cost because of the expensive equipment needed to drill the borehole.

#### *4.4.3. Simulation models of GHEs*

The main objective of the GHE thermal analysis is to determine the temperature of the heat carried fluid, which is circulated in the U-tube and the heat pump, under certain operating conditions. Actually, the heat transfer process in a GHE involves a number of uncertain factors, such as the ground thermal properties, the ground-water flow rate and building loads over a long lifespan of several or even tens of years. In this case, the heat transfer process is rather complicated and must be treated, on the whole, as a transient one. In view of the complication of this problem and its long time scale, the heat transfer process may usually be analysed in two separated regions [3]. The heat transfer models for the two separate regions are as follows.

- *Heat conduction outside borehole*. A number of simulation models for the heat transfer outside the borehole have been recently reported, most of which were based on either analytical methodologies or numerical methods [3].

- Kelvin's line-source. The earliest approach to calculating the thermal transport around a heat exchange pipe in the ground is the Kelvin's line-source theory, i.e. the infinite line source [20]. According to the Kelvin's line-source theory, the temperature response in the ground due to a constant heat rate is given by:

$$t(r,\tau) - t_0 = \frac{q}{4\pi\lambda}\int_{\frac{r^2}{4a\tau}}^{\infty} \frac{e^{-u^2}}{u} du \tag{23}$$

where $r$ is the distance from the line-source and $\tau$ the time since start of the operation; $t$ is the temperature of the ground at distance $r$ and time $\tau$; $t_0$ is the initial temperature of the ground; $q$ is the heating rate per length of the line source; $\lambda$ and $a$ are the thermal conductivity and diffusivity of the ground. The solution to the integral term in Eq. (23) can be found from the related references [21].

- Cylindrical source model. The cylindrical source solution for a constant heat transfer rate was developed by Carslaw and Jaeger [22], then refined by Ingersoll et al. [21], and later employed in a number of research studies [23]. Based on the governing equation of the transient heat conduction along with the given boundary and initial conditions, the temperature distribution of the ground can be given in the cylindrical coordinate:

$$\begin{aligned} &\frac{\partial^2 t}{\partial r^2}+\frac{1}{r}\frac{\partial t}{\partial r}=\frac{1}{a}\frac{\partial t}{\partial \tau} && r_b<r<\infty \\ &-2\pi r_b\lambda\frac{\partial t}{\partial \tau}=q && r=r_b,\ \tau>0 \\ &t-t_0=0 && \tau=0,\ r>r_b \end{aligned} \tag{24}$$

where $r_b$ is the borehole radius.

The cylindrical source solution is given as follows:

$$t-t_0=\frac{q}{\lambda}G(z,p) \tag{25}$$

where $z=a\tau/r_b$, $p=r/r_b$.

As defined by Carslaw and Jaeger [22], the expression $G(z, p)$ is only a function of time and distance from the borehole centre. An approximate method for $G$ was proposed by Hellström [19].

- Eskilson's model. Both the one-dimensional model of the Kelvin's theory and the cylindrical source model neglect the axial heat flow along the borehole depth. A major progress was made by Eskilson [24] to account for the finite length of the borehole. The basic formulation of the ground temperature is governed by the heat conduction equation in cylindrical coordinates:

$$\begin{aligned} &\frac{\partial^2 t}{\partial r^2}+\frac{1}{r}\frac{\partial t}{\partial r}+\frac{\partial^2 t}{\partial z^2}=\frac{1}{a}\frac{\partial t}{\partial \tau} \\ &t(r,0,\tau)=t_0 \\ &t(r,z,0)=t_0 \\ &q(\tau)=\frac{1}{L}\int_D^{D+L} 2\pi r\lambda\left.\frac{\partial t}{\partial r}\right|_{r=r_b} dz \end{aligned} \tag{26}$$

where $L$ is the borehole length; $D$ means the uppermost part of the borehole, which can be thermally neglected in engineering practice.

In Eskilson's model, the numerical finite-difference method is used on a radial–axial coordinate system to obtain the temperature distribution of a single borehole with finite length. The final expression of the temperature response at the borehole wall to a unit step heat pulse is a function of $\tau/\tau_s$ and $r_b/L$ only:

$$t_b-t_0=-\frac{q}{2\pi\lambda}f(\tau/\tau_s,r_b/L) \tag{27}$$

where $\tau_s = L^2/9a$ means the steady-state time. The *f*-function is essentially the dimensionless temperature response at the borehole wall, which was computed numerically.

- Finite line-source solution. Based on the Eskilson's model, an analytical solution to the finite line source has been developed by a research group which considers the influences of the finite length of the borehole and the ground surface as a boundary [3]. The solution of the temperature excess was given by Zeng et al. [25]:

$$t(r,z,\tau) - t_0 = \frac{q}{4\pi\lambda}\int_0^L \left[\frac{erfc\left(\dfrac{\sqrt{r^2+(z-l)^2}}{2\sqrt{a\tau}}\right)}{\sqrt{r^2+(z-l)^2}} - \frac{erfc\left(\dfrac{\sqrt{r^2+(z+l)^2}}{2\sqrt{a\tau}}\right)}{\sqrt{r^2+(z+l)^2}}\right]dl \tag{28}$$

It can be seen from Eq. (28) that the temperature on the borehole wall, where $r = r_b$, varies with time and borehole length. The temperature at the middle of the borehole length ($z = L/2$) is usually chosen as its representative temperature. An alternative is the integral mean temperature along the borehole length, which may be determined by numerical integration of Eq. (28).

- *Heat transfer inside borehole*. The thermal resistance inside the borehole, which is primarily determined by thermal properties of the grouting materials and the arrangement of flow channels of the borehole, has a significant impact on the GHE performance. The main objective of this analysis is to determine the entering and leaving temperatures of the circulating fluid in the borehole according to the borehole wall temperature, its heat flow, and the thermal resistance [3].
- One-dimensional model. A simplified one-dimensional model has been recommended for GHE design, which considers the U-tube as a single "equivalent" pipe [26].
- Two-dimensional model. Hellström [19] derived the analytical two-dimensional solutions of the thermal resistances among pipes in the cross-section perpendicular to the borehole axis, which is superior to empirical expressions and one-dimensional model.
- Quasi-three-dimensional model. On the basis of the two-dimensional model mentioned above, a quasi-three-dimensional model was proposed by Zeng et al. [27], which takes account of the fluid temperature variation along the borehole depth.

#### 4.4.4. *Ground thermal response test*

In the case of vertical closed-loop GCHP systems, the determination of the parameters to calculate the vaporization thermal power that must be provided from the ground is laborious. To know how many loops must be set, which is a function of the energy that must be given to the heat pump, evaluating the thermal conductivity of the ground and the effective thermal resistance of the borehole are very important. In this respect, taking a thermal response test (TRT) of the ground is necessary, using a borehole in which a simple ground loop is placed.

During an in-situ test, a ground electric heater usually provides heat to the circulating fluid (water or glycol) through the ground loop while the inlet ($t_i$) and outlet ($t_e$) fluid temperatures are measured (Figure 9). The average of these two instantaneous temperature reading is usually taken to represent the average temperature in the vertical ground loop at a given time. In an ideal test, the measured circulating flow rate and the heat input rate remain constant throughout the test [28]. The first TRT in Romania was performed in 2009 by the GEOTHERM PDC company of Bucharest [29].

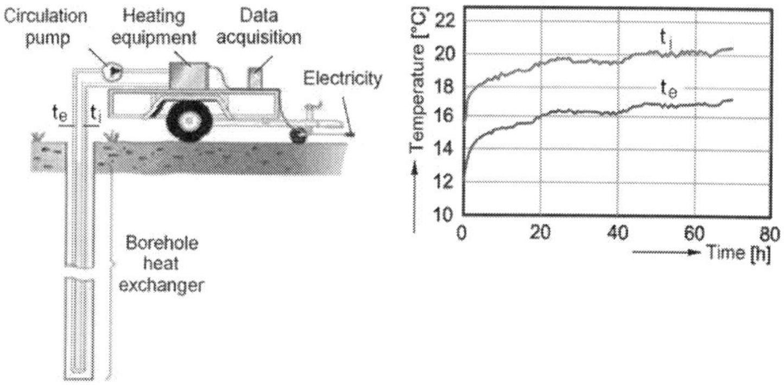


**Figure 9.** Schematic of equipment for thermal response test

To estimate the minimum duration $\tau_{min}$ of the test, the following equation can be used [24]:

$$\tau_{\min} = \frac{5r_b^2}{a} \tag{29}$$

where $r_b$ is the borehole radius and $a$ is the ground thermal diffusivity.

For data analysis and final evaluation of ground thermal conductivity $\lambda$ and borehole thermal resistance $R_b$, some methods were developed [15,30] that use one of the simulation models of GHEs previously presented. Through the ground thermal response test, the length of the borehole is properly determined, the operating performance of the system is provided, and supplementary costs (extra loops, boreholes, glycol, etc.) are avoided. This operation is performed using specialized software.

## 5. Heat pump heating and cooling systems

### 5.1. Radiator heating system

A hot-water radiator heating system is a type of central heating. In the system, heat is generated in the boiler. For the generation of the heat, a natural gas boiler is used where the chemical energy of natural gas is transferred into the heat. Then, the heat is distributed by hot-water (heat carrier) to the radiators. The radiators heat the rooms. The hot-water is circulated by a water circulation pump, which operates continuously. The radiators, as rule of thumb, are located next to the cold surfaces of the envelope. They significantly influence the thermal comfort. The radiators release the highest amount of heat to the heated room by convection and one part by heat radiation [31]. The convective heat transfer will lead to a lower relative humidity of the air, and, at high radiator surface temperature, dust particles can be burned, leading to lower indoor air quality. Thus, emitters should be implemented with a radiation factor as high as possible in the case of high-temperature water supplies. The highlights of the convective thermal field achieved with radiators were illustrated in [32].

To ensure ever-changing heat demand in a room, qualitative, quantitative or mixed control systems are used.

### 5.2. Radiant heating and cooling systems

In low-energy buildings, the low-temperature heating system usually works with a supply water temperature below 45°C [33]. Embedded radiant systems are used in all types of buildings. Radiant heating application is classified as panel heating if the panel surface temperature is below 150°C [34]. In thermal radiation, heat is transferred by electromagnetic waves that travel in straight lines and can be reflected. The water temperatures are operated at very close to room temperature and, depending on the position of the piping, the system can take advantage of the thermal storage capacity of the building structure.

Figure 10 shows the available types of embedded hydronic radiant systems [35]. Panel heating provides a comfortable environment by controlling surface temperatures and minimising air motion within a space. A radiant system is a sensible heating system that provides more than 50% of the total heat flux by thermal radiation. The controlled temperature surfaces may be in the floor, walls, or ceiling, with the temperature maintained by circulation of water or air. The radiant heat transfer is, in all cases, 5.5 W/(m$^2$K). The convective heat transfer then varies between 0.5 and 5.5 W/(m$^2$K), depending on the surface type and on heating or cooling mode. This shows that the radiant heat transfer varies between 50 and 90% of the total heat transfer [36].

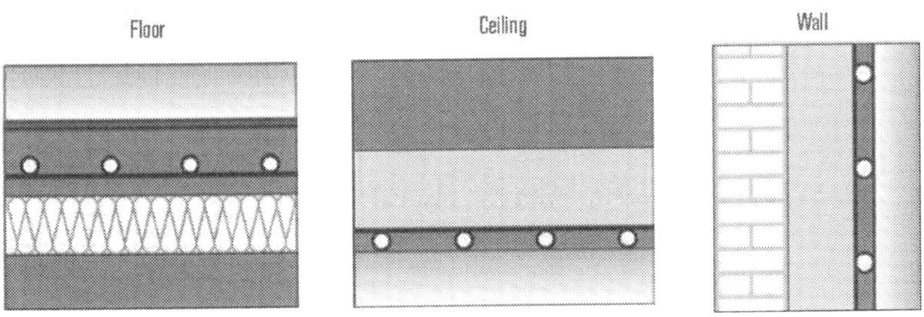


**Figure 10.** Examples of water-based radiant systems

Radiant panel heating is characterised by the fact that heating is associated with a yielding of heat with low temperature because of physiological reasons. Thus, at the radiant floor panels, the temperature must not exceed +29°C, and at the radiant ceiling panels the temperature will not exceed 35–40°C, depending on the position of the occupier (in feet) and the occupier distance to the panels, in accordance with thermal comfort criteria established by ISO Standard 7730 [37]. For cooling, the minimum floor temperature is 19°C. A vertical air temperature difference between head and feet of less than 3°C is recommended.

### 5.3. Performance assessment of radiator and radiant floor heating systems connected to a GCHP

#### 5.3.1. *Description of office room*

Experimental investigations of GCHP performance were conducted in an office room (Figure 11) at the Polytechnic University of Timisoara, Romania, located at the ground floor of the Civil Engineering Faculty building. The Timisoara city has a continental temperature climate with four different seasons. The heating season runs in Timisoara from 1 October to 30 April. The following data are known: heat transfer resistance (1/*U*-value) of building components: walls (2.10 m²K/W), ceiling (0.34 m²K/W), windows and doors (0.65 m²K/W); glass walls surface, 8.2 m²; total internal heat gain (e.g. from computers, human and lights), 25 W/m²; and heat demand, 1.35 kW. The indoor and outdoor air design temperatures are 22 and –15°C, respectively.

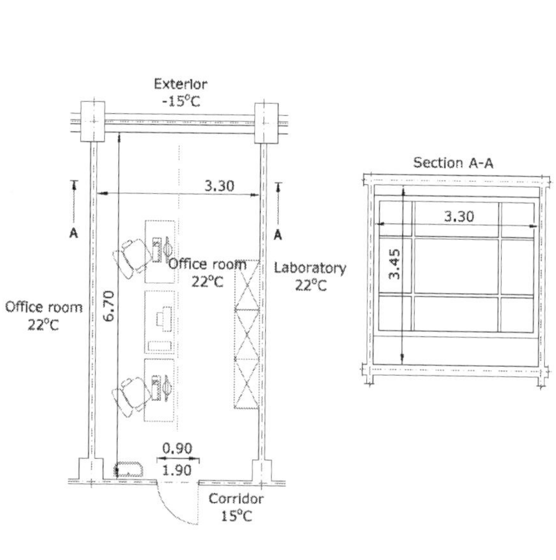


**Figure 11.** Heated office room

This space is equipped both with a floor heating system and steel panel radiators to analyse the energy and environmental performances of these systems. These two heating systems are

Chapter 2

# Geothermal Energy as a Major Source of Renewable Energy - Learning from Asian Neighbours

Ram Avtar and Pankaj Kumar

Additional information is available at the end of the chapter

http://dx.doi.org/10.5772/62084

**Abstract**

Energy is one of the fundamental necessities for economic growth, increased social equity, and an environment that allows livelihoods to thrive. Development is not possible without energy, and sustainable development is not possible without sustainable energy. Renewable Energy (RE) is a key to sustainable development and offers an opportunity to improve access to modern energy services for the impoverished segments of society, which is essential for sustainable community development. In Asia, Indonesia and the Philippines are rich in Geothermal (GT) energy. Being a renewable, sustainable, and indigenous energy resource, GT is reliable because its supply is independent of season and global energy market dynamics. Japan has a huge potential for GT and wants to expand renewable energy resources after the Fukushima Nuclear Power Accident. In the near future, GT is set to play a major role in national energy supply in Japan. Therefore, the objective of this study is to look at various challenges and opportunities in GT energy policies of Indonesia and the Philippines to adopt them to meet Japanese needs in the future.

**Keywords:** Geothermal energy, Energy potential, Policies

## 1. Introduction

Global population has been increasing exponentially and we are estimating that between the years 2000 and 2050, the population will grow from 6.1 billion to 9.6 billion (United Nations 2012). The average energy consumption per person between years 1850 and 2010 has increased from about 4.8 x 109 to over 77.3 x 109 J/person/year. This means the average person is using more than 15 times the energy each year than they used previously [1]. Sustainable development goals adopted in September, 2015 calls for use of sustainable energy systems as one of its main agenda. Among all sustainable options available, development and optimization for the procurement of geothermal energy is one of the best because it is produced and used in a way that is compatible with the well-being of future generations and the environment [2]. Geothermal energy is one of the most significant alternate sources of energy, with a much higher output of electricity production per MW of capacity as compared to wind or solar. Most of the countries with GT resources should promote themselves to explore it with the technical advancement and social acceptance because it has various advantages like round-the-clock availability, independent of time and climatic conditions. This means that to generate electricity at a constant level, geothermal energy can be supplied as base load in contrast to hydro or wind power. Geothermal energy's potential is ubiquitous, environmentally friendly, and only marginally developed (International Geothermal Association). In general, global warming or climate change is not expected to have any substantial impacts on success rate of geothermal energy utilization. However, the wise implementation and inclusion of geothermal energy in our energy demand mix could play a meaningful role in mitigating climate change. Best suitable site for commercial production of geothermal energy are high-temperature (>180°C) hydrothermal systems associated with recent volcanic activity and located near active plate tectonic boundaries (i.e., convergence zones, subduction zones, transform faulting), or at crustal and mantle hot spot anomalies. On the other hand, regions with high radioactive isotope decay, which increases terrestrial heat flow or where aquifers are charged by water heated through circulation along deeply penetrating fault zones, cannot be neglected as potential sites to harness geothermal energy if managed efficiently.

The United States has the world's largest geothermal energy potential (39,325 MW) followed by Indonesia (27,791 MW), and the Philippines has about 6,000 MW, while the target country, Japan, has a potential of 22,540 MW (METI, 2010). However, geothermal energy generation requires a very high initial cost and involves a high development risk, which hinders its growth. There is a visible gap between the countries that have policies supporting deregulated energy markets and the ones that do not. Japan's energy policy has its shortcomings when it comes to enumerating specific mechanisms to support increased participation from the private sector and other stakeholders to promote inclusive growth and fuel greater investments in geothermal energy development. Also, use of geothermal energy for applications in agriculture, tourism, etc., and the policy aspects for the same need to be included. These low and medium enthalpy applications will take place at the community level and need specific inputs regarding the involvement of small households and other stakeholders in the process of development.

In Indonesia and the Philippines, increased investments, especially for research and exploration for geothermal energy development come from greater private sector partnerships, which are fuelled through a proactive energy policy pursued by the respective governments. Indonesia, under its "Roadmap for Geothermal power," pursues a policy of management of geothermal working areas (GWAs) by private entities, selected through a tender process and provides specific fiscal incentives for geothermal sector, like income tax facilities and customs duty exemption. The policy provides opportunities wherein the private sector can use CERs (Certified Emission Reductions) to generate carbon revenue. In the Philippines, for the past four decades, geothermal energy has been developed by fostering government and private sector cooperation. In 1990, the Philippine government passed the BOT (Build, Operate, Transfer) law to invite private sector investment, which helped to launch financing schemes with private contractors. The EPIRA (Electric Power Industry Reform Law) of 2001, led to the liberalization of the power industry. All these policy instruments have significantly contributed to the capacity addition in geothermal energy sector, with a possibility of offsetting the high costs involved in geothermal energy development. These successful cases will be used to review existing lacunae in the Japanese approach.

The policy frameworks related to the financing of geothermal energy and its applications will be reviewed in the proposed chapter, by looking at the existing practices in Japan, and comparing it from successful policies practiced in Indonesia and the Philippines. A review of the existing "feed-in tariff system" for renewable energy will be done, and suggestions for specific inclusions, regarding geothermal energy will be made, as currently the policy recognizes "other renewable energy sources" as cost-prohibitive. A thorough review of the mechanisms used for deregulation, especially in the selected cases of Indonesia and the Philippines will be done, to analyze the shortcomings and successes of the various policies like EPIRA. A criterion for implementation of these policies in the local Japanese context will be done. Also, a review of the processes utilized for policy implementation will be done, looking at the effectiveness of certain policy instruments. Another crucial aspect will be to assess community-based development of direct uses of geothermal energy, an area which has not been explored adequately. Heating, drying, agriculture, and tourism sector policies will be explored and cases of similar implementation in Indonesia and the Philippines will be looked at, to build policy capacity and to engage with stakeholders at different levels.

United Nations and its member countries have adopted sustainable development goals (SDGs) in September, 2015, where they sought for a total 17 goals for sustainable livelihood of the world community. Goal 7 clearly emphasizes on providing access to affordable, reliable, sustainable, and modern energy for all. Here every country is looking to increase substantially the share of renewable energy in the global energy mix and double the global rate of improvement in energy efficiency, enhance international cooperation to facilitate access to clean energy research and technology, expand infrastructure, and upgrade technology for supplying modern and sustainable energy services for all in developing countries, in particular least-developed countries, small island developing States, and landlocked developing countries, in accordance with their respective programs of support by the year 2030. In a nutshell, this goal is sought for potential renewable energy resource like geothermal. As population is increasing in geometric proportions and so is the demand for resources, it is the need of the hour to think diligently and act seriously to trap energy sources like geothermal, which is with us just around the corner. It is important to keep striving for better technology to harness this precious source of energy efficiently.

## 2. Energy scenarios

### 2.1. Current scenario in Japan

Japan ranks fifth in the world for the amount of annual energy consumed (following China, United States, Russia, and India), and third for fossil fuel consumption including oil and natural gas (BP, 2013). Also from the environmental perspectives, it would need to reduce GHG emissions and in 2013 Japanese government announced that Japan will achieve -3.8% targets as compared to 2005 by 2020. Considering the fact that Japan does not have fossil fuel deposit and only 25–30% of total energy demand is being supplied at domestic level (nuclear and renewable sources), it relies heavily on import of fossil fuels from other countries like Indonesia. In total, renewable energy contribution to total energy production is approximately 5% (with hydro being main contributor with 3.3%) in Japan, which is projected to increase to 20% by 2020 to become more energy independent as well as to overcome energy security challenges.

However, for achieving such a goal, Japan has to scrutinize and revaluate all the challenges and opportunities present in its geopolitical scenarios. Geographically speaking, Japan is considered to be one of the potential countries for geothermal energy because of its location within the Pacific "ring of Fire", and this potential is considered to be one of the largest in the world (Figure 1).

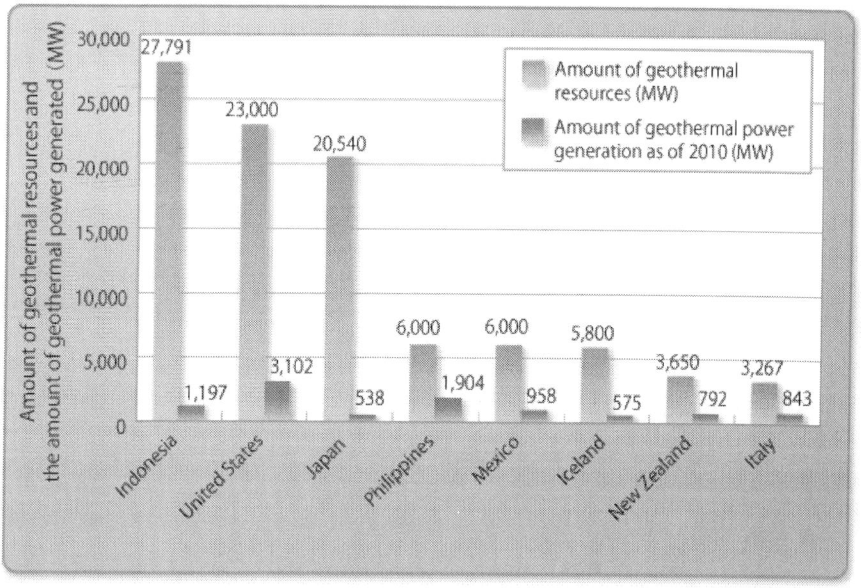


**Figure 1.** Amount of geothermal resources available worldwide and the amount of geothermal power generated (Source: http://www.asiabiomass.jp/english/topics/1208_04.html)

Before detailed feasibility analysis of technological advancement for geothermal energy harvesting, it is worth looking at the current status of this non renewable energy source in neighboring countries Indonesia and the Philippines.

It is observed that there is an increase in the direct use of geothermal energy from 58% to 78% in a decade (2000–2010) based on the data from 82 countries [4]. They mention that approximately 55% is used for ground-source heat pumps, followed by 20% for bathing and swimming (including balneology), 15.0% for space heating, 4.5% for greenhouses and open ground heating, 2.0% for aquaculture pond and raceway heating, 1.8% for industrial process heating, 0.4% for snow melting and cooling, and 0.4% for agricultural drying [4].

### 2.2. Status of geothermal energy in Indonesia

Indonesia is the largest archipelago in the world. It consists of over 17,000 islands and approximately 70,000 villages. Meanwhile, because of the large territory, 45% of the villages out of 70,000 are located in remote areas and about 6,200 villages have not been supplied by electricity. The population of Indonesia in 2010 reached 237 million, or is increasing by average of 1.48% per year. Indonesia's steady economic growth of more than 6%, even during the recent global recession, was accompanied by a 9% growth in electricity demand each year [5,6]. The increase of energy consumption is derived from the sectors of industry, transportation, and commercial. Total current annual primary energy demand in Indonesia is 168.7 MTOE (million tons of oil equivalents), which is predominantly fuelled by oil followed by coal, biomass, and gas, and supports the country's economy. It is reported that the ratio of production reserves for coal, gas, and oil in Indonesia are going to run out in 75 years, 33 years, and 12 years, respectively, so opting for renewable energy sources, especially untapped vast source of geothermal energy, is the viable solution to sustain its economic growth [7]. The current use of fossil fuel covers up to 32% of the share for total energy demand, which is expected to increase up to 42%, if Indonesia adopts MP3EI (Master Plan for the Acceleration and Expansion of Indonesian's Economic Development) scenarios. Since 2004, Indonesia withdrew its membership from OPEC (Organization of the Petroleum Exporting Countries) as it became a net importer of oil. Indonesia's huge geothermal potential is facing risks and challenges to meet the target of development as expected, while Indonesia as non-Annex-1 country is committed to reduce GHG emission from 26% up to 41% by 2020.

Indonesia's dependency on fossil fuels leads to two unwanted consequences – a strained government budget, while simultaneously undermining the country's climate change mitigation efforts. The fuel subsidy to ensure the availability of cheap energy (in order to achieve political mileage) has turned into a huge fiscal burden for the state, amounting to nearly 21% of total government expenditure in 2005, and continues to rise [8]. On the other hand, excess use of fossil fuels is also projected to raise Indonesia's greenhouse gas (GHG) emissions fourfold by 2030 [9]. This growth in emissions is in contradiction to the commitments made by Indonesia's President to reduce GHG emissions by 26% by 2020 and to increase the use of renewable energy so that it accounts for 25% of total energy production by 2025 [10, 11, 12].

### 2.3. Indonesia's energy security

Dependency on imported oil does not look promising in terms of long run of viable economy, although there is a potential for other energy resources such as geothermal. Even energy mix policy for the country led to high dependency on oil in the recent past. A cumulative impact of economic slowdown, increasing rate of fuel, and increasing energy subsidy results in high economic instability. In general, subsidies tend to cause overconsumption of the resource, since the market price does not reflect the actual cost of producing one unit of petroleum product.

They also discourage energy efficiency measures and the development of alternative or renewable energy sources by way of low electricity tariffs. The state budget is heavily burdened by this policy and in order to provide low-priced electricity, access is denied to nearly half the population. "This policy mostly favors the urban population or those who are privileged enough to have access to electricity while forgoing the development of necessary new infrastructure needed to deliver electricity to those without it" [8]. Against the backdrop of increasing fossil fuel prices, as well as growing concerns over energy security, renewable energy is well-positioned to play a critical role in Indonesia's energy mix. By expanding the use of geothermal energy in particular, the government has within its means another tool to better manage its economic future. Use of geothermal energy can ease high dependency on oil and it will consequently reduce the burden from fossil fuel and electricity subsidy [13].

Currently Indonesia practices fixed type of Feed-in Tariff (FIT) fund, set by the government to support geothermal projects that have won tenders but cannot continue development as they need a tariff above the fixed amount. The FIT fund helps in the payment of the difference between the price required by geothermal developers and the current electricity price. Indonesia introduced the concept of a ceiling price (9.7 USD/KWH) in 2009, below which the winning tender bid would automatically be accepted, but above which the bid was subject to negotiation with local government. With this model, the risks now lie with the local governments.

Geothermal energy production is capital-intensive, and will require billions of dollars in debt and equity to realize the targets set by government. The cost of exploration alone, for the next 3,000 MW is estimated at 2.8 billion USD. With little of this likely to come from government, the bulk of this will need to be raised from the local private sector and donor agencies. In principle, there is no shortage of potential sources of concessionary finance: the World Bank Group, Asian Development Bank, IFI, JICA, and JBIC have ambitious plans to provide support, including CTF funds. Still, the limitation here is no visibility about the hidden cost of geothermal energy generation (cradle-to-grave concept), so it should not be limited to estimates of the levelled cost of energy, but also take into account the likely financial costs. For instance, one should consider cost for risk of non recovery, grid and storage technology, and political barriers. One of the most fundamental obstacles to reaching the government target is the fact that most geothermal resources are located in remote areas, which means additional financing is required to connect electricity production to the main grid [14].

Another source of uncertainty, which translates to risk, is the confusion when it comes to division of power between central and local governments. With decentralization, regional governments are important because they become the official owners of the steam resource, whereas the central government plays an equally pivotal role providing expertise and underwriting the power purchase agreements [15]. While fine in theory, decentralization seems to have only raised transaction costs. Apart from institutional challenges, i.e., bureaucratic delay and no proper coordination between different ministries only leads to delay in executing all projects on time and thus further financial burden. Low purchase price of electricity by the state-owned power companies further deters the progress of geothermal power generation in Indonesia (Asia biomass office).

The other institutional challenge is the involvement of ministry of finance and forestry. Ministry of finance is engaged in driving financial incentives crucial to attracting investors into the sector. Furthermore, the ministry, through its sovereign wealth investment agency, plays an important role in creating publicly funded initiatives designed to mobilize private capital, such as the Clean Technology Fund [16]. On the other hand, The Ministry of Forestry also plays a crucial role in geothermal energy development as most identified reservoirs lie beneath protected forest areas. With some explicit policy support for accelerating geothermal energy development, it is imperative that measures are taken so that lands acquired for geothermal energy use are not high conservation value forests or sensitive ecosystems, and that the impacts and risks on forests are mitigated. To increase the geothermal energy production there is need to develop a cooperative and cross-sectoral approach to minimize the conflicts and make a progressive move toward the development of geothermal energy resources.

Another major concern about development of geothermal energy production system is the mind-set of local authority. Activities related to geothermal energy production is considered as mining activities, resulting in prohibition of use of protected/conserved areas like forest area. Local government or nongovernmental agencies should make an effort to classify the geothermal activities as extraction of thermal energy, which is different from mining activities of coal or other mineral resources with potential to lead to deterioration of environmental essence.

Last but not the least, some other issues not listed above are lack of technique/data to estimate the actual potential of geothermal resources availability to generate more interest among investors, no lucrative price of per unit of energy, limited equity funds and uncertainty in legal aspects, and the lack of cross-sector coordination.

Finally, it can be said that Indonesia has huge geothermal resources, but to date arrangement (technical and other setups) can only harness approximately 5% of the potential. Up to September, 2014, only 1396 MWe or approximately 4.5% of the potential has been installed. There is a need to harmonize and to synchronize the regulation among difference stakeholders, different government and financial institutions, etc. Indonesia also requires support on capacity building, in order to increase expertise and also to recruit human resources ready to support geothermal development to realize its goals.

### 2.4. Status of geothermal energy in the Philippines

The Philippines is an archipelago of 7,107 islands surrounded by South-East Asia's main bodies of water. The population continues to grow at 1.5% annually, reaching 135.2 million by 2035. This rapidly developing nation is certain to face significant challenges to energy security if it intends to maintain the current course of its economic development over the next few decades. The total current annual primary energy demand for the Philippines is around 39.4 MTOE (million tons of oil equivalents). About 59% of it is being supplemented by fossil fuel, while 40% is supported by renewable energy. The Philippines, however, imports most of its oil and is likely to continue to do so to sustain the economy's total petroleum requirements of 27.9 MTOE by 2035. The largest consumer of this energy is the transportation sector (35%), followed by residential and industrial sectors, roughly the same, accounting for about 26% each. The commercial sector constitutes about 11.9% and agriculture, forestry with the smallest portion is 1.3%. To provide energy security, although indigenous energy generation in the Philippines is quite modest and accounts for 22% from geothermal, 12% from biomass, and 6% from hydropower, it still has to cover a large distance to minimize the dependency on imported oil. As the major instrument for realizing the energy sector's vision of achieving energy independence, the Department of Energy (DOE) is currently crafting the 2012–2030 Philippine Energy Plan (PEP). As embodied in the Electricity Power Industry Reform Act (EPIRA), the economy's electricity supply industry has been restructured, paving the way for the privatization of the state-owned National Power Corporation (NPC). While oil pricing is deregulated, electricity pricing is a regulated energy commodity. The price for electricity is set by the Energy Regulatory Commission (ERC). Alongside the implementation of the EPIRA is the unbundling of electricity rates. In view of the Philippines' wide-ranging geographical situation, to fully connect the entire population to the national grid is a significant hurdle. Servicing the most remote and difficult-to-electrify rural areas will require significant resources; hence, achieving a 100% electrification level over the outlook period remains a challenge for the economy. Among all the other types of renewable energy sources like hydro and wind power there are lots of constraints because of geoclimatic factors, which ultimately leads to further push to geothermal energy as potential source of energy.

### 2.5. Solution common for both countries

#### *2.5.1. Energy mix policy*

Sustainable energy policies are likely to succeed if they also contribute toward other societal and economic development objectives in the future. From the standpoint of economic analysis, the optimum quantity of geothermal energy that should be in energy mix is given by intersection of the geothermal supply curve with the avoided social costs of thermal energy.

#### *2.5.2. Feed-in Tariff*

Should tariff ceilings be based on estimates of production costs, or on the basis of estimated benefits?

A tariff should be rational, and in support of clearly defined objectives. This would ensure that the resources are not developed for their own sake simply because they exist, and because it is generally held to be desirable. The tariff methodology should be transparent (and documented as part of a tariff issuance), with clearly stated assumptions.

- A tariff should promote economic efficiency. A tariff should be adaptable to changing circumstances. This requires the methodology to have a defined basis and provide for review and updating to a clearly stated timetable.
- Stakeholders should be consulted. While consensus is not always achievable, concerns should be addressed.

FIT is generally understood as being based on the production costs of the technology in question, as in the original German model. The distinguishing feature of FITs is that they are technology specific, and often differentiated by project size and other technical characteristics (such as additional bonus payable for projects that meet criteria for domestic content or other technical attributes seen as desirable). Competitively determined tariffs are the best guarantee of economic efficiency. It is sometimes assumed that tariff ceilings should be based on today's costs (if based on production costs), or today's benefits (if based on PLN's avoided costs). But that is not reasonable if the intent is to apply the ceilings to a tender for a project that will deliver benefits only 6–8 years from the tender date. It should include the cost of non recovery of energy resources, establishing grading and storage technology after that period of time from tender date.

*2.5.3. Measures to manage social issues*

i. **Conduct "free, prior and informed consent" public consultation:** "Free, prior and informed consent" (FPIC) is the principle that an indigenous community has the right to give or withhold its consent to proposed projects that may affect the lands they customarily own, occupy, or otherwise use.

ii. **Implement awareness and acceptance programs:** Prior to any discussion about a project, the company/developer must introduce itself to its various stakeholders, consisting of the local government units (LGUs), government agencies, host communities, nongovernmental organizations (NGOs), peoples' organizations (POs), and private business.

iii. **Create a multistakeholder monitoring team:** Ensure that mechanisms are in place so that project activities can be monitored by a Multisectoral Monitoring Team (MSMT) composed of representatives from the local government units, host community, NGOs, the Department of Environment and Natural Resources (DENR), and other concerned sectors in the area.

iv. **Set up an environmental guarantee fund:** This is a financial arrangement negotiated between the proponent, the government, and the affected community. The amount is intended for rehabilitation and payment of damages due to the accidents from the operation of the project.

v. **Provide economic packages:** Social acceptability is often equated with the stakeholders' access to meaningful benefits or benefits that have direct positive impacts. These need to be shared equally with communities in recognition of their contribution to national security and national development for hosting the project.

vi. **Resettle dislocated communities, if necessary**

vii. **Protect prior and ancestral rights:** Ancestral domain shall be fully recognized and protected by the project.

viii. **Protect forest patrimony:** Ensure that the project will not disrupt forest patrimony.

*2.5.4. Funding the large project*

Recently, most of the geothermal projects have received a positive nod from UNEP's Sustainable Energy Finance Initiative and Bloomberg's New Energy Finance for big financial help in response to high oil prices and the threat of climate change.

### 2.6. Challenges and opportunities for geothermal energy: A road map for Japan

The Philippines and Indonesia are classified as non-Annex-I countries in the United Nations Framework Convention on Climate Change (UNFCCC), but Japan is different as it comes under the list of Annex-I countries. Countries like the Philippines and Indonesia have adopted emissions reduction targets under the Copenhagen Accord, and have adopted national policies and strategies for climate change adaptation and mitigation. But for Japan, international pressure to reduce the GHGs emission and per capita energy consumption rate is mounting. One of the obstacles against geothermal energy development in Japan is the fact that most of the promising fields are located near or inside national parks or spa resorts. Recently, the Ministry of Environment of Japan allowed the extraction of geothermal energy from national parks that could potentially be developed into a big renewable energy source and reduce its dependence on nuclear and other resources. Looking at the neighboring countries as stated in section above and being prone to natural disasters (especially Fukushima nuclear disaster, 2011), Japan should look for safe and potential energy source, particularly geothermal energy. So far, photovoltaic-cells-based solar energy has been the most dominant renewable energy source, mainly because it can be easily deployed on rooftops of existing buildings and houses and connected to grids in the form of distributed power supplies requiring much less effort and cost than that for geothermal energy. The adoption of renewable energy in general, visible in the electricity source mix, has grown gradually, but it is still less than 2% of overall electricity. Since the government is the key stakeholder in energy business, this is certainly a great advocacy move for renewable energy to grow further in the Japanese market by following institutional strengthening factors:

- Continue favorable Feed-in Tariff (FIT) as fundamental policy
- Develop better energy mix policy
- Shorten the time taken for environmental assessment in developing renewable energy sites
- Address and solve current problems like high costs for electricity generation, intermittent energy supplies, restriction of development locations, through feasible long-term and short-term policy measures

Looking for small-to-medium-scaled geothermal power aligned with technological advances such as low cost binary cycles utilizing existing hot springs and heat pumps utilizing temperatures differences just beneath the surface of ground, both of which do not require a huge risky investment such as drilling the new ground very deeply, do not take years for addressing the impact on surrounding nature and for exploration, and can be implemented without a huge distribution plant will be a good source as primary energy because they can continuously provide stable sustainable energy source throughout a year regardless of changes in weather and no ill-effect on the grid stability.

Finally, apart from the challenges, optimization of geothermal energy usage will lead to all win-win situation for all around the society by giving positive social impacts (reducing poverty, enhancing equality, health, community safety), being environmentally benevolent (no environmental pollution as well as biodiversity conservation) and renewable, efficiently produced, country like Japan must have a big go to these projects.

## Author details

Ram Avtar* and Pankaj Kumar

*Address all correspondence to: avtar@unu.edu

United Nations University, Institute for the Advanced Study of Sustainability (UNU-IAS), Tokyo, Japan

## References

[1] Ingvar, B., Fridleifsson, Ruggero Bertani, R., Ernst Huenges, E., John, W. L.und, Arni Ragnarsson, A., Ladislaus Rybach, L. 2008.The possible role and contribution of geothermal energy to the mitigation of climate change. IPCC Scoping Meeting on Renewable Energy Sources – Proceedings, 59-80.

[2] Shortall, R., Davidsdottir, B., Axelsson, G. 2015. Geothermal energy for sustainable development: A review of sustainability impacts and assessment frameworks. *Renew Sustain Energy Rev*, 44: 391-406.

[3] Asia biomass office report: Indonesia is Doubling the Purchase Price of Electricity from Geothermal Power Generation to Accelerate its Adoption (http://www.asiabiomass.jp/english/topics/1208_04.html)

[4] John, W. L.,und, Tonya, L. B.oyd, 2015. Direct Utilization of Geothermal Energy 2015 Worldwide Review. Proceedings World Geothermal Congress 2015, Melbourne, Australia, 19-25 April, 2015.

[5] Ministry of Finance. 2009. Ministry of Finance Green Paper: Economic and Fiscal Policy Strategies for Climate Change Mitigation in Indonesia, Ministry of Finance & Australia Indonesia Partnership, Jakarta.

[6] PwC (PricewaterhouseCoopers). 2011. Electricity in Indonesia – Investment and Taxation Guide, PwC, Jakarta, viewed 2 January 2012, at (http://www.pwc.com/id/en/publications/assets/Indonesian-Electricity-Guide_2011.pdf).

[7] Nasruddin, M., Alhamid, I., Daud, Y., Surachman, A., Sugiyono, A., Aditya, H.B., Mahila, T.M.I. 2016. Potential of geothermal energy for electricity generation in Indonesia: A review. *Renew Sustain Energy Rev*, 53, 733-740.

[8] Resosudarmo, B.P., Alisjahbana, A., Nurdianto, D.A. 2010., "Energy Security in Indonesia, ANU Working Papers in Trade and Development No. 2010/08, Canberra.

[9] Fiscal Policy Office of the Ministry of Finance) & World Bank. 2011. "Phase 1 assessment of emissions – key findings," viewed 16 March 2011, at (http://www.esmap.org/esmap/sites/esmap.org/files/factsheet04.pdf). IGA-International geothermal association, http://www.geothermal-energy.org

[10] Fadillah, R.D. 2011. "More German companies eye renewable energy sector," *The Jakarta Post*, 25 October, viewed 16 March 2011, at (http://www.thejakartapost.com/news/2011/10/25/moregerman-companies-eye-renewable-energy-sector.html).

[11] President of the Republic of Indonesia. 2011. "Peraturan Presiden no. 61 tahun 2011 tentang Rencana Aksi Nasional Penurunan Emisi Gas Rumah Kaca" (Presidential Decree No. 61 of 2011 on National Action Plan for the Reduction of Greenhouse Gas Emissions), The Secretary of the Cabinet of the Republic of Indonesia, Jakarta.

[12] The Secretary of the Cabinet of the Republic of Indonesia 2011. "Cabinet Secretary: President Yudhoyono signed presidential decree on national action plan to reduce GHG emissions," Press Release, 26 September 2011, the Secretary of Cabinet of the Republic of Indonesia, Jakarta.

[13] WWF., 2012. Igniting the ring of fire: A vision for developing Indonesia's geothermal power.

[14] Tanoto, Y., Wijaya, M.E. 2011. Economic and Environmental Emissions Analysis in Indonesian Electricity Expansion Planning: Low-rank Coal and Geothermal Energy Utilization Scenarios, *International Journal of Renewable Energy Research*, Vol. 1, 2011.

[15] de Wilde, A. 2010. Accelerating Geothermal Development in Indonesia. Jakarta.

[16] The World Bank Energy Sector Management Assistance Program. 2010. Geothermal handbook: planning and financing geothermal power generation. Washington DC: The World Bank. 2010.

Chapter 3

# Radiogenic Heat Generation in Western Australia — Implications for Geothermal Energy

Mike F. Middleton

Additional information is available at the end of the chapter

http://dx.doi.org/10.5772/61963

**Abstract**

The chapter reviews heat generation in crystalline rocks and influences on overlying sedimentary basins in Western Australia (WA). Regions of elevated thorium and uranium will cause elevated heat generation, which in turn can cause elevated heat flow. Western Australia hosts several large sedimentary basins with the potential for hot sedimentary aquifers (HSAs). These include the Perth, Carnarvon, and Canning basins. Parts of these basins are underlain by crystalline rocks that contain high levels of heat-generating elements, such as uranium, thorium, and potassium. Also, the Pilbara Craton, which contains both sedimentary and crystalline rocks, that entertains a number of active mines, which may benefit from geothermal energy, is investigated. Further, the southern part of the Perth Basin (Vasse Shelf), which is underlain by crystalline rocks with elevated concentrations of thorium and uranium, is shown to possess higher than usual temperatures. From observations, and geothermal modeling, it is concluded that the Perth Basin has a high potential for medium- to low-temperature geothermal energy developments. In other parts of Western Australia, the Carnarvon Basin has elevated temperatures in artesian groundwater. Heat flow in the Canning Basin is briefly reviewed; this basin has some geothermal potential, but it is far from the major population centers.

**Keywords:** Radiogenic heat, Uranium, Thorium, Potassium, heat flow

## 1. Introduction

The information presented in this study reflects the view in Western Australia to move toward mechanisms for sustainable energy into the future. The Western Australian legislative framework supports this view, and continuous studies have been carried out since 2008 to implement this vision into the future. The initial formal Western Australian governmental view was to use geothermal energy for electricity generation, and legislation was formulated to accommodate this. By late 2014, it became apparent that geothermal energy in Western Australia was taking a different path. The chapter focuses, both formally published works and less formal studies, on radiogenic heat generation within rocks in Western Australia and their contribution to heat flow, elevated temperatures, and geothermal energy potential. It adopts the approach that heat flow from basement (essentially crystalline) rocks beneath sedimentary basins will elevate temperatures within the sedimentary basin, and thus create a natural reservoir of hot fluids in the sediments that can be used for geothermal energy purposes. These are referred to as hot sedimentary aquifers.

The connection between radioactive elements in the earth and heat generation has been known for many years. Among many, the following workers have developed this science [1–3], and in specific reference to Australia [4–7]. About this period, an increasing awareness was growing for the potential need for geothermal energy in the Australian energy mix, and a landmark book was published by Beardsmore and Cull [7]. Studies were undertaken by Geoscience Australia, the national geoscience survey organization in Canberra [8–10]. In Western Australia, one of the significant early workers to recognize that geothermal energy may be viable in the state was by Bestow [11]. In his study, Bestow [11] recognized parallels of Western Australian basins to the Paris Basin and the need to achieve temperatures in the vicinity of 65–85 °C. At this early stage, Bestow also clearly recognized that the geothermal potential in Western Australia was best based around the concept of hot sedimentary aquifers, rather than either the engineered geothermal systems (EGSs) being advocated elsewhere in Australia [12,13] or the more commonly exploited volcanogenic heat sources.

Early work on heat flow and radiogenic heat production was carried out by Jaeger [4], Sass et al. [5], Middleton [6], Cull and Denham [14], and Cull [15]. These studies identified the background to understanding heat flow and heat generation in Western Australia. Perhaps rather more by serendipity than by intention, the study by Jaeger [4] identified a rather unique site in the "Wheat Belt" of Western Australia (Figure 1). This site was in a granite quarry near the small town of Doodlakine, and returned a heat generation value of about 21.9 heat generation units (hgu; 1 hgu = 0.418 $\mu W\ m^{-3}$) at the surface and 21.2 hgu as an average of 30 samples from an associated bore hole. This translates to 9.15 $\mu W\ m^{-3}$ in SI units. At the time, this was the largest heat generation value published for a Western Australian granitoid (a term used herein to refer to a felsic igneous rock with largely granitic mineralogy). More recent work on heat generation in Western Australia was carried out by Middleton [16], Middleton and Stevens [17], and Middleton et al. [18]. During these studies, the Doodlakine site was revisited and now initially reported; surface measurements of heat generation of the granite near the original measurements published by Jaeger [4] were made, using an RS-125 gamma-ray scintillometer (see below for description of the technique). Twelve measurements were made on the granite outcrop, near where Jaeger (1970) carried out his investigation, and these yielded a mean heat generation of 8.5 $\mu W\ m^{-3}$ (20.3 hgu used by Jaeger [4]). This is a good agreement, given slightly different assumptions made for conversion of uranium (U), thorium (Th), and potassium (K) concentrations to heat generation, made by Jaeger [4] as compared with Middleton [16].

http://dx.doi.org/10.5772/61963

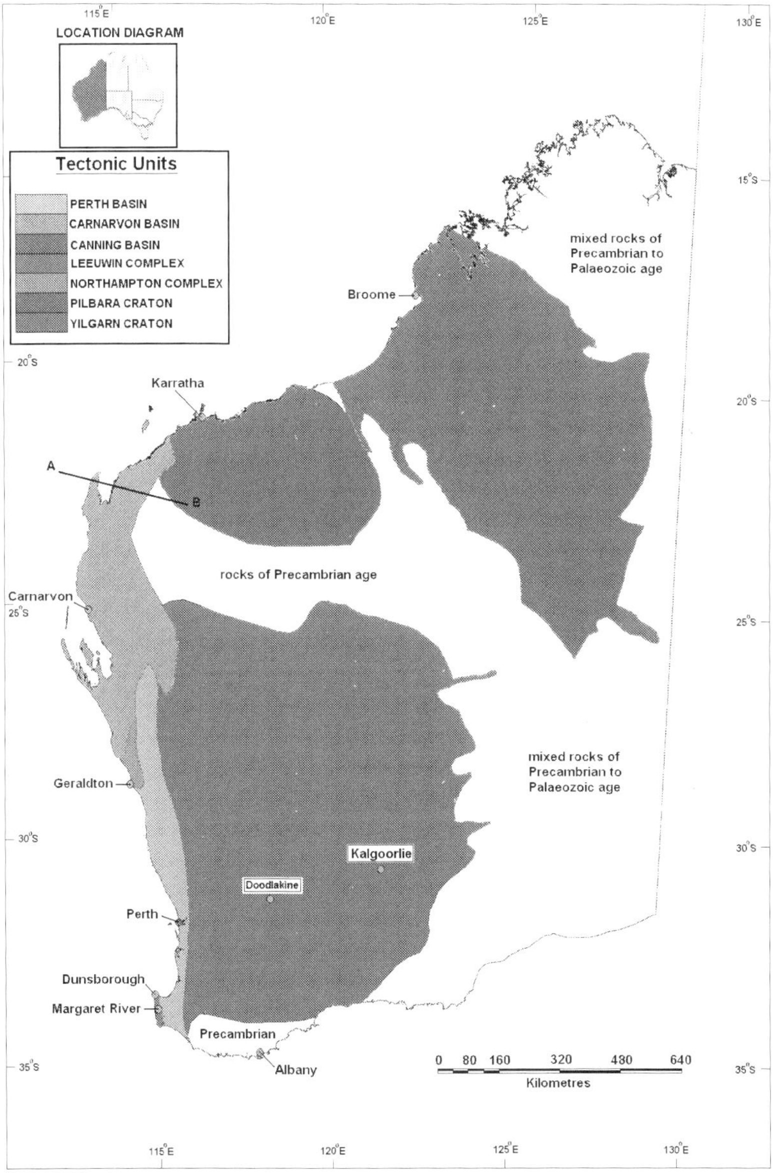


**Figure 1.** WA map showing tectonic provinces. Note that the Doodlakine location is halfway between Perth and Kalgoorlie.

Uranium distribution, as with Th distribution, in Western Australia is poorly known at depth. However, a reasonable knowledge of surface U distribution is known from airborne radiometric data compiled in the "uranium merge" and "thorium merge" maps of the state produced by the Geological Survey of Western Australia [19]. The U and Th merge maps are shown in Figure 2. It must be recognized that these maps have limitations because of the attenuation of radiation, due to many effects, for example, the distance from surface radioactive source rocks to the sensor in the aircraft, time of day of acquisition, the limitations (because of different instrumentation and calibration) in the merge process for various survey datasets, and the difference in surface geology across the state [20,21]. From the maps, it is observed that Th concentration (parts per million, ppm) is generally higher than the U concentration. Thorium to uranium ratio (Th/U) has been reviewed by Middleton et al. [18] for various surface locations (Table 1). It should be noted that the surface U and Th concentrations from airborne measurements are commonly lower than surface measurements of concentration, because of the environmental effects commonly observed in deriving U and Th from the surface "merge" maps [20,21]. From the data in Table 1, the Th/U ratio showed a variation between 2.0 and 13.0 for outcrop rocks. Uranium and thorium are often concentrated in deeply emplaced felsic igneous (granitoid) rocks. However, these elements can also be concentrated in sedimentary rocks. In recognizing this, some surface measurements of U and Th in sedimentary basins may be higher than normal due to sedimentary depositional and enrichment processes.

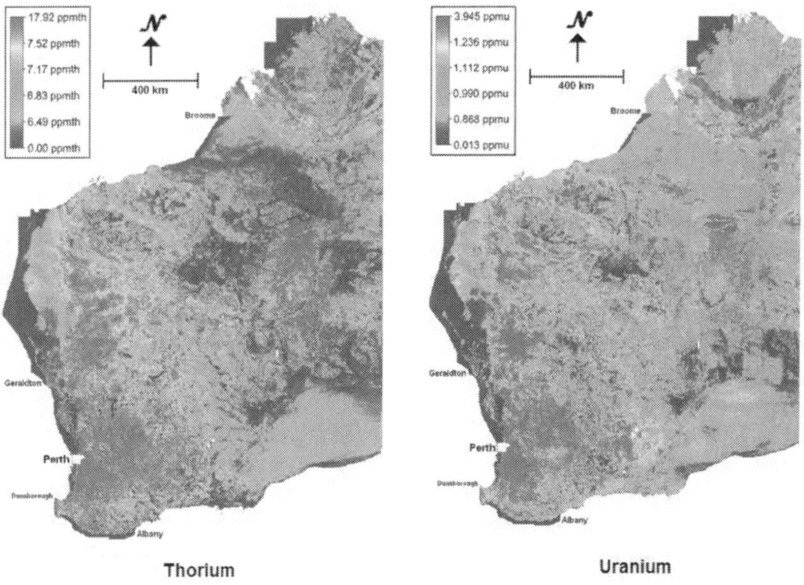


**Figure 2.** Map with surface U and Th surface distribution in WA. These are based on "merge maps" produced by Brett [19] for the Geological Survey of Western Australia.

http://dx.doi.org/10.5772/61963

| LOCATION | µR/hr | $A_0$ (µW/m³) | Th (ppm) | U (ppm) | K(%) | Comment |
|---|---|---|---|---|---|---|
| Sugarloaf 1 | 480.5 | 9.9 | 108.4 | 6.0 | 7.4 | Leeuwin (granitoid) |
| Sugarloaf 2 | 130.3 | 2.2 | 21.3 | 1.3 | 3.8 | Leeuwin (granitoid) |
| Gracetown N | 251.0 | 4.2 | 49.1 | 0.7 | 6.3 | Leeuwin (granitoid) |
| Gracetown S | 154.7 | 2.6 | 24.4 | 1.8 | 4.6 | Leeuwin (granitoid) |
| Leeuwin LH | 217.7 | 3.9 | 37.8 | 2.8 | 5.6 | Leeuwin (granitoid) |
| Wagon Wheel | 257.0 | 4.8 | 46.8 | 3.3 | 6.2 | Leeuwin (granitoid) |
| Meelup Pk 1 | 600.8 | 15.7 | 168.0 | 12.6 | 6.2 | Leeuwin (granitoid) |
| Meelup Pk 2 | 485.3 | 12.3 | 140.9 | 7.6 | 4.9 | Leeuwin (granitoid) |
| Meelup Pk 3 | 843.0 | 40.1 | 178.9 | 32.3 | 6.3 | Leeuwin (enrichment) |
| Point Picouet | 147.1 | 2.2 | 20.4 | 0.1 | 7.0 | Leeuwin (granitoid) |
| Cowaramup | 389.3 | 11.5 | 82.5 | 20.3 | 4.4 | Leeuwin (granitoid) |
| Farquhar Rd | 574.0 | 15.2 | 147.2 | 18.7 | 0.4 | S. Perth B (sediment) |
| Carbunup 1 | 59.5 | 1.6 | 13.3 | 2.6 | 0.2 | S. Perth B (sediment) |
| Carbunup 2 | 49.0 | 1.4 | 10.1 | 2.7 | 0.1 | S. Perth B (sediment) |
| Rosa Brook 1 | 479 | 12.0 | 129.7 | 10.7 | 0.1 | S. Perth B (enrichment) |
| Rosa Brook 2 | 198 | 4.9 | 56.2 | 3.7 | 0.02 | S. Perth B (enrichment) |
| Stoneville | 200.1 | 4.5 | 32.6 | 6.9 | 3.9 | Darling R (granitoid) |
| Ashenden Rd | 241.5 | 5.4 | 44.7 | 7.3 | 3.8 | Darling R (granitoid) |
| Golden View 1 | 242.6 | 5.9 | 36.8 | 11.0 | 4.1 | Darling R (granitoid) |
| Golden View 2 | 255.7 | 5.7 | 42.1 | 9.0 | 4.5 | Darling R (granitoid) |
| Darlington | 138.6 | 3.3 | 20.7 | 6.1 | 2.5 | Darling R (granitoid) |
| Chalet Rigi | 369.5 | 10.2 | 55.7 | 22.7 | 3.8 | Darling R (granitoid) |
| J Forrest Pk1 | 309.8 | 8.0 | 45.0 | 17.0 | 4.0 | Darling R (granitoid) |
| J Forrest Pk 2 | 245.3 | 6.0 | 42.6 | 10.1 | 3.5 | Darling R (granitoid) |
| J Forrest Pk 3 | 285.5 | 7.3 | 43.2 | 14.7 | 4.2 | Darling R (granitoid) |
| J Forrest Pk 4 | 279.0 | 7.3 | 40.1 | 15.8 | 3.7 | Darling R (granitoid) |
| Parkerville | 224.3 | 5.8 | 42.2 | 9.6 | 3.8 | Darling R (granitoid) |
| Canning Dam | 176.3 | 4.3 | 30.2 | 7.3 | 2.6 | Darling R (granitoid) |
| Mundaring 1 | 3.6 | 0.9 | 7.6 | 1.5 | 0.2 | Darling R (sediment) |
| Mundaring 2 | 4.1 | 1.3 | 8.1 | 2.1 | 1.9 | Darling R (sediment) |
| Mundaring 3 | 25.1 | 6.4 | 68.6 | 6.0 | 0.2 | Darling R (laterite) |
| Doodlakine | 359.0 | 8.2 | 76.7 | 10.4 | 4.0 | Central Yilgarn (gr.) |
| Cunderdin | 297.6 | 6.3 | 65.2 | 5.0 | 4.5 | Central Yilgarn (gr.) |
| Goongarrie | 7.5 | 1.1 | 7.7 | 1.9 | 2.6 | Central Yilgarn (gr.) |

**Table 1.** Western Australian RS-125 measurement campaign for U, Th, K & $A_0$ for the years 2011 and 2015.

Potassium usually has a different genesis to U and Th. It is a very commonly occurring element in the earth's crust. The abundance (concentration) of K in most igneous rocks is commonly observed to be about 2–6%, and it also occurs abundantly in many sedimentary rocks. Beardsmore and Cull [7] have discussed the specific abundance of the radioactive isotope (K-40) within normal potassium distributions in rocks. The surface K "merge" distribution map [19] can be misleading, as the K concentrations tend to follow recent sedimentary depositional features, such as seasonally dry riverbeds. In such deposits, the time of year of that an airborne radiometric survey was acquired has implications for interpretation of the observed K concentration [20–22]. Further, as will be seen in the following sections, the contribution of K to radiogenic heat generation is usually less than the contribution of U and Th [7,16,23].

In this review, surface-based geophysical or laboratory measurements of U, Th, and K are emphasized rather than airborne measurements. This is because surface-based (on-the-ground) measurements of outcropping rocks tend to be more directly indicative of the particular rock type being investigated, rather than a "bulk" estimate of the chemistry of the rocks being sampled by an airborne survey. This will be discussed further below.

## 2. Theory

This section reviews the basic, but well-known, concepts to understand heat generation, heat flow, and simple temperature distributions with the earth's crust.

### 2.1. Heat flow and temperature due to heat generation in a radiogenic layer

Heat flow at the surface on the earth ($Q_s$) over a series of $n$ layers with internal heat generation and stable tectonic environment can be determined by the equation

$$Q_s = Q_b + A_1H_1 + A_2H_2 + \dots + A_nH_n, \tag{1}$$

where $Q_b$ is the heat flow from the base of the $n$ layers being considered, typically the upper mantle, $A_1$ is the heat generation in layer 1, $H_1$ is the thickness of layer 1, $A_2$ is the heat generation in layer 2, $H_2$ is the thickness of layer 2, $A_n$ is the heat generation in layer $n$, and $H_n$ is the thickness of layer $n$.

Temperature within a surface layer of thickness $L$, and uniform heat generation of $A_0$, is shown by Carslaw and Jaeger [24], described by the equation:

$$T = \left[\frac{A_0 z}{2K}\right](2L - z) + \frac{Q_b z}{K} + T_s, \tag{2}$$

where $T$ is the temperature, $z$ is the depth of observation, $K$ is the thermal conductivity, $T_s$ is the mean annual surface temperature, and $Q_b$ is the heat flow at the base of the layer. The geometry applicable to this equation is shown in Figure 3, and this equation can be applied to estimate the temperature versus depth in a granite batholith with uniform radiogenic heat generation.

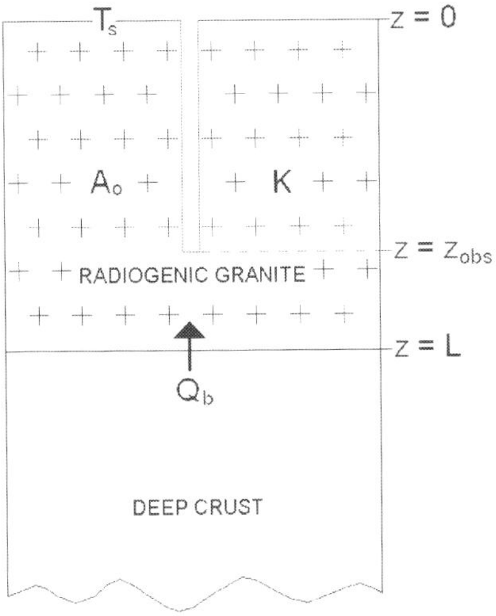


**Figure 3.** theory for temperature model. Note that the same modeling geometry is used for relatively thin enrichment layer models in section 7. In the latter case, the radiogenic granite layer is a layer where U- and Th-rich fluids have been rapidly emplaced due to chemical alteration processes, largely due to weathering and erosion.

The more general time-variable equation describing the gradual rise in temperature due to thermal adjustment of the emplacement of a layer of thickness $L$ at an initial time of $t$=0 is given by Carslaw and Jaeger [24] for $0 < z < L$:

$$T = \left[\frac{\alpha A_0 t}{K}\right]\left[1 - 4i^2\text{erfc}\left(\frac{z}{\tau}\right) + 2i^2\text{erfc}\left(\frac{L+z}{\tau}\right) - 2i^2\text{erfc}\left(\frac{L-z}{\tau}\right)\right] + T_0\text{erf}\left(\frac{z}{\tau}\right) + \frac{Q_b z}{K}, \tag{3}$$

where $T$ is the temperature, $z$ is the depth, $t$ is the time, $\alpha$ is the thermal diffusivity, $K$ is the thermal conductivity, $A_0$ is the heat generation, $T_0$ is the initial temperature of layer at $t = 0$, $Q_b$ is the heat flow at the base of layer, erf($x$) is the error function, erfc($x$) is the complimentary error function [1 – erf($x$)], $i^2$erfc($x$) is the second integration of the error function, and $\tau = (4\alpha t)^{½}$. The temperature modeling below is largely based on these three equations. A theoretical case, based on Eq. (3), where a radiogenically enriched layer of variable thickness occurs, by chemical alteration processes, at the earth's surface is also investigated in Section 7.

The case of sedimentary layers overlying a radiogenic granite is also important. This situation has the potential to cause a "thermal blanketing effect," where the overlying sediments possess significantly lower thermal conductivity than the underlying rocks. In many cases, this has been shown to create anomalously high temperatures within the upper sedimentary sequence. Such cases have been treated theoretically by Carslaw and Jaeger [24]. In the Western Australian perspective, the "thermal blanketing" effect was investigated in the case of a geothermal (temperature) anomaly in the northern Perth Basin (Gillingarra anomaly) and is discussed below. This investigation showed that the wavelength of the observed heat flow anomaly was too low (and magnitude too high) to accommodate a radiogenic heat source from an estimated 10-km basement depth.

### 2.2. Heat generation from U, Th, and K

Studies of radiogenic heat generation from the decay of unstable isotopes of U, Th, and K have been summarized by Kappelmeyer and Haenel [23], Jessop [25], Beardsmore and Cull [7], and Middleton [16]. The generalized relation [16] used for the present study is

$$A_0 = (0.26)\ (\text{U ppm}) + (0.07)\ (\text{Th ppm}) + (0.1)\ (\text{K \%}), \tag{4}$$

where $A_0$ is the heat generation, (U ppm) is the concentration of uranium in ppm, (Th ppm) is the concentration of thorium in ppm, and (K %) is the concentration of potassium in percent. The relationship used in Eq. (4) is consistent with that of Beardsmore and Cull [7]. It is recognized that by using these constants in Eq. (4), the density of all rocks is assumed to be a constant value of 2700 kg $m^{-3}$ (as is similarly assumed in the example of Beardsmore and Cull [7]). Essentially, this assumption represents a generalized conversion formula.

### 2.3. Field measurements with an RS-125 gamma-ray spectrometer

For studies reviewed in this chapter, many of the sites measured for U, Th, and K assay were obtained using an RS-125 gamma-ray spectrometer. The RS-125 instrument is calibrated, via "assay mode", to provide an approximation of U, Th, and K concentrations in a half-space being sampled. The manufacturer indicates that an instrumental error of about 10% is incurred in the RS-125 in the assay mode, due to the programmed calibration matrix. The volume sampled is assumed to be a perfect half-space, and variations from this perfect half-space can also cause some errors in the assay determination. However, observation locations were chosen so as to minimize this type of error. The manufacturer's specifications of the RS-125 spectrometer indicate that the instrument reflects a sample area of a radius of approximately 1 m in the half-space (Figure 3).

Calibration procedures were carried out prior to, and during, the field acquisition with the RS-125. An important factor for calibration is the sample time of each measurement. In determining this, the manufacturer asserts that the calibration matrix, which is used to optimize the spectral definition of the three isotopes, is designed to provide the minimum error (< 5%) for sample times of approximately 4 minutes. However, this has to be balanced against the requirement for long sample times (often over many hours) for rocks with low U, Th, and K concentrations, as expected in many of the granites being investigated. It was finally decided to base all measurements for studies reported in this review to be 5 minutes [16]. Further, the RS-125 field assays were compared with U, Th, and K assays obtained by laboratory measurements for sample sites in the Leeuwin Complex in the southwest of Western Australia, and good correspondence was observed (Figure 4).

**Figure 4.** A diagram showing the RS-125 sampling zone, which is essentially a 1-m-radius hemisphere below the earth's surface.

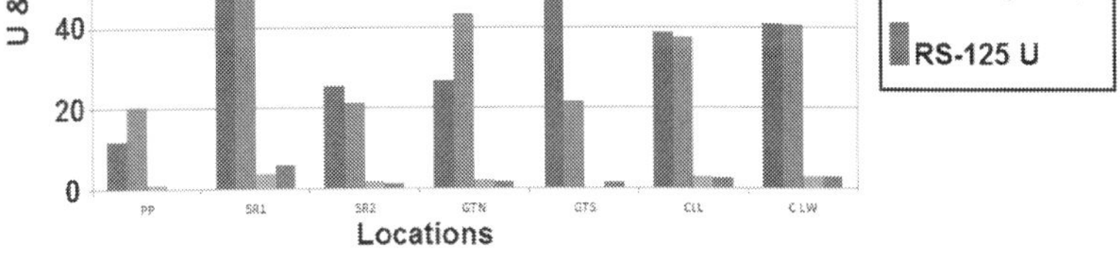


**Figure 5.** Correlation of RS-125 to laboratory assays for Leeuwin Complex samples.

### 2.4. Heat generation derived from gamma logs

Ryback [26] and Bücker and Ryback [27] proposed empirical equations to determine heat generation ($A_0$, $\mu Wm^{-3}$) from wire-line gamma-ray (GR) logs run in (petroleum and sometimes mineral) exploration wells or boreholes. The Bücker and Ryback [27] correlation appears to be superior to the earlier correlation by Ryback [26], and is expressed as

$$A_0 = 0.0158\left(\text{GR}\right) - 0.0126, \tag{5}$$

where $A_0$ is the heat generation (in units of $\mu Wm^{-3}$) and GR is the logged gamma-ray value in API (American Petroleum Institute) units. A gamma-ray API is defined, and further discussed in relation to other radioactivity measurement units, in Society of Petroleum Engineers (SPE) [28]. When looking at gamma-ray measurements greater than 20 API units, this equation simplifies to $A_0 \approx 0.0158$ (GR). For this study, it is assumed that $A_0 = 0.0158$ (GR).

Beardsmore and Cull [7] investigated this relationship with regard to the East Yeeda 1 well from the Canning Basin in Western Australia. They derived a relationship between heat generation and the gamma-ray and density logs from this well, which had total count and spectral gamma-ray logs. These workers derived a relationship between the heat generation and the two wire-line-log values to be

$$A_0 \approx 0.005\left(\text{GR}\right)\left(\text{RHOB}\right), \tag{6}$$

where RHOB (g $cm^{-3}$) is the bulk density and GR (API units) are the readings from the wire-line log. The authors suggest that an uncertainty in $A_0$ is about ± 0.5 $\mu W\ m^{-3}$. These workers place the caveat that this relationship may be local correlation, rather than a general relationship. If we assume a general rock density of 2700 kg $m^{-3}$, then the Beardsmore and Cull relationship becomes

$$A_0 \approx 0.014\left(\text{GR}\right). \tag{7}$$

On the basis of these considerations, Eqs (5) and (7), heat generation may be in a very general sense correlated to the wire-line log gamma-ray value by a relationship represented by the range 0.014 (GR) < $A_0$ < 0.016 (GR), with GR in API units and $A_0$ in $\mu W\ m^{-3}$.

From 2011 to 2015, a series of studies was carried out with a commercial gamma-ray spectral scintillation detector, RS-125 instrument, as described below. This instrument was used to measure many of the heat generation estimates reported in this study [16,18]. It is of importance to establish how this instrument performs, and provides results, consistent with other radioactive measuring instruments. Since 2011, the results shown in Figure 6 have been accumulated, which can provide an approximation of dose rate (DR; as reported by the instrument) with respect to uranium (ppm), thorium (ppm), and potassium (K). This correlation can provide a correlation of dose rate to heat generation via Eq. 4. Correlations, according to the data in Figure 6, can be established between DR and $A_0$, with heat generation values ($A_0$) being determined according to Eq. (4). Regression curves have been established for (1) DR total and (2) DR <300 μR/hr. It is believed that DR (μR/hr) < 300 may be more appropriate for correlation studies, because of the nonlinearity of the DR versus $A_0$ relationship above the value of 300 μR/hr. In the case of all the data being incorporated, a second-degree polynomial curve DR = 0.6292 $(A_0)^2$ + 47.186 $(A_0)$, with regression coefficient of $R^2$ = 0.9254, is found to provide the best match (Figure 6).

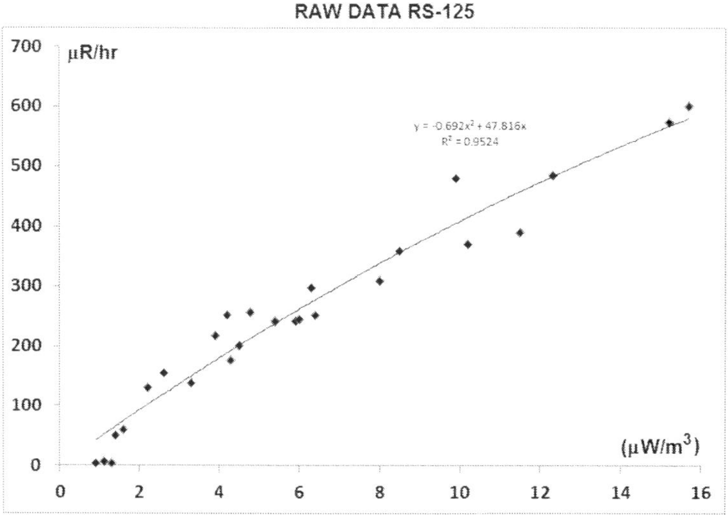


**Figure 6.** Correlation between RS-125 dose rate in (üR/hr) and heat generation (μW/m³) from the data derived by field measurements [16,17].

In the special case of DR being < 300 μR/hr, a regression match is found to be DR = 43.21 $(A_0)$, with a regression coefficient of $R^2$ = 0.8866. In this case, an intercept at zero is also assumed. The observed data for $A_0$ less than about 1.3 $\mu Wm^{-3}$ (DR < about 15 μR/hr) appear to be showing a limitation in the calibration of the scintillation crystal to accurately measure the uranium, thorium, and potassium elemental concentration spectra, and more research is required to look into why this is happening. Despite the concerns of the calibration of the RS-125 crystal at low U, Th, and K concentrations, the author believes, that sufficient confidence can be placed in the data for DR >15 μR/hr, that an acceptable correlation between DR and $A_0$ can be sustained to be DR = 0.43.21 $(A_0)$. Assuming the SPE correlation between DR and API gamma-ray measurements is DR = 0.67 API units [28], the data found by Middleton [16] indicate a correlation of DR ≈ 0.0155 (GR). This correlation is very close to the Bücker and Ryback [27] approximation.

It is proposed that an approximate conversion factor for GR (API units) to $A_0$ ($\mu Wm^{-3}$ units) for this investigation to be

$$A_0 = 0.015\left(\text{GR}\right). \tag{8}$$

This equation is the mean of the Bücker and Ryback [27] and Beardsmore and Cull [7] approximations. Some estimates of heat generation, derived from this equation, of the gamma-ray response of basement rocks observed in wire-line logs from the Perth Basins are shown below.

## 3. Perth basin

The Perth Basin lies in the southwest corner of Western Australia (Figure 1). It contains sediments of Phanerozoic age, which are surrounded by crystalline granitoid and metamorphic rocks of the Yilgarn Craton, Leeuwin Complex, and Northampton Complex (Figure 1). The Perth Basin will be treated in three parts: the southern Perth Basin [29], the central Perth Basin [30–32], and the northern Perth Basin [33].

### 3.1. Southern perth basin

Heat generation in the granites of the Leeuwin Block, which immediately crops out to the west of the Vasse Shelf, may have relevance for the exploration for geothermal energy "hot spots" in the Vasse Shelf sediments (Figure 1). The significance of measuring heat generation in the Leeuwin Complex is that it is expected that these rocks underlie the elements of the Perth Basin immediately to the east of these outcropping sediments. The elements of the Perth Basin in question are the Vasse Shelf sediments (largely of Permian age) and the deeper and thicker Bunbury Trough sediments (largely of Cretaceous to Permian age) [29]. It is uncertain if the Leeuwin Complex crystalline rocks extend under the Bunbury Trough, which has been described geologically by Crostella and Backhouse [29].

Western Australia has several regions where hot granitoids are known to occur in outcrop. The locations of high radioactivity rocks, including hot granitoids, can initially be detected by airborne radiometric surveys. These data have been acquired for Western Australia through a number of cooperative projects between the Geological Survey of Western Australia and Geoscience Australia, and maps showing these data are publically available. The Vasse region was flown in 2011. The U concentration map for the northern part of the Vasse Region, derived from this survey, is shown in Figure 2a, and the Th map for the same region is shown in Figure 2b. A ground-truthing survey was carried out by the members of the Petroleum Division of the Department of Mines and Petroleum in May 2013. This survey used a RS-125 gamma-ray spectrometer to obtain surface values of U, Th, and K to compare with the airborne data. Middleton [16,34] carried out a similar study to investigate geothermal heat generation for the Darling Range granitoids. Surface sampling of granitoids of the Leeuwin Complex was previously carried out by the Geological Survey of Western Australia and Geoscience Australia, and these sites were also revisited during 2011 and 2012 to compare the spectrometer values to those obtained by the previous laboratory-based measurements of U, Th, and K (main locations shown in Figure 8). The U, Th, and K concentrations in the granitoids can be converted into approximate radiogenic heat generation using Eq. (4); see Figure 7.

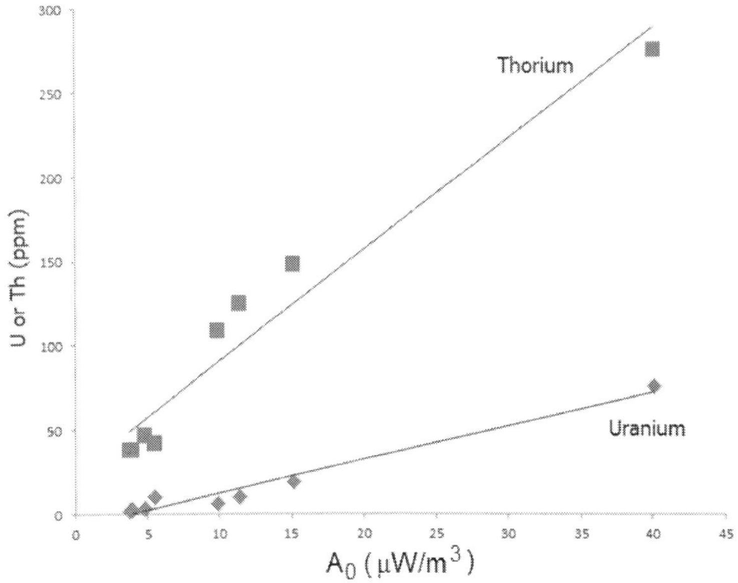


**Figure 7.** Correlation of U or Th concentration to heat generation, using Eq. (4).

Previous works [4,5,16,34,35] have been carried out on heat generation in granitoids of the western part of the Yilgarn Craton. The heat generation of outcropping hot granites in the Yilgarn Craton has been measured in the range of 1–10 μWm$^{-3}$ and is similar to published values for the Cooper Basin Hot Dry Rock (HDR) granites (ca. 10 μWm$^{-3}$) [6]. Some previous heat generation values were published for the Leeuwin Complex by Middleton and Stevens [17], although U, Th, and K analyses have been measured by laboratory techniques from outcrop samples for selected locations (see Figure 7). More recent heat generation values have been acquired in the northern part of the Leeuwin Complex and are herein reported.

Table 1 shows the mean U, Th, and K concentrations (measured with at least three repeats) at the 16 localities in this Leeuwin and south Perth Basin reviewed in this study. The elemental abundances observed at these sites were converted to heat generation in units of μWm$^{-3}$ using the factors in Eq. (4). Table 1 shows the relationship between U, Th, K and heat generation (designated as $A_0$). The table indicates that the observed mean heat generation at the locations investigated ranged between 3.8 μWm$^{-3}$ and 40.1 μWm$^{-3}$ throughout the Leeuwin Complex region. The heat generation data exhibit a good linear relationship between U and Th concentrations. The very high U and Th concentrations, and calculated heat generation of 40.1 μWm$^{-3}$, at the Meelup Park anomaly location (MP in Figure 8) near Dunsborough are indeed anomalous, which was investigated because of the high airborne radiometric signals (see Figures 8–10). The reason may be due to the concentration of these elements in a thin laterite or gravel layer residing on the top of a hot underlying granite body. No obvious outcrop of the underlying granitic rocks was observed during the survey at this locality.

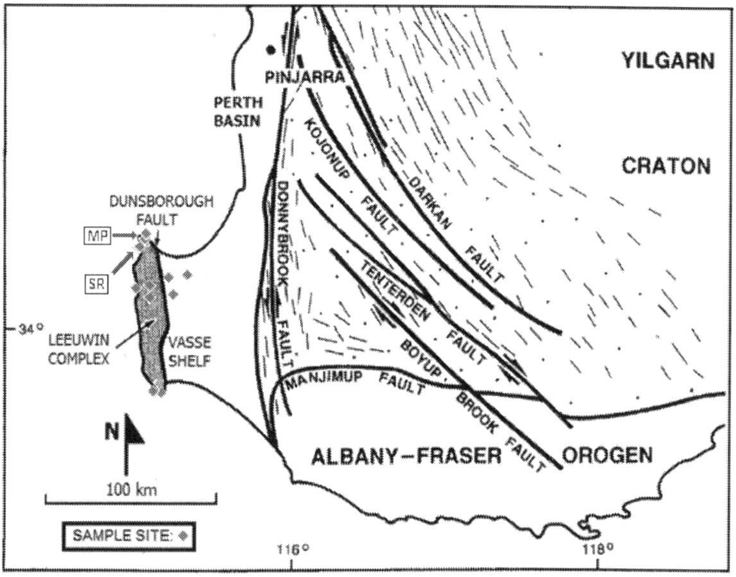


**Figure 8.** South Perth Basin structural and other tectonic elements.

The airborne signature suggests that the underlying hot granite may have an areal extent of about 5 km$^2$. Figure 11 shows the thorium anomalies superimposed on the aeromagnetic data at the Meelup Park location. The anomalous thorium appears to be located in a low magnetic region in the range of −50 to −200 nT (green to light blue in Figure11). The low magnitude of the magnetic signature of the thorium-rich regions implies that the thorium may be located in low magnetic crystalline rocks. This tends to suggest that the thorium (and uranium) may be associated with the more felsic rocks in the Leeuwin Complex, which is also supported by evidence from the Sugarloaf Rock locality. The Sugarloaf Rock locality is located on the coast about 3 km south of Cape Naturaliste (see Figure 8). Laboratory measurements by the Geological Survey of Western Australia indicate at least two igneous rock suites at the locality: (a) with U ~3.9 ppm and Th ~92.6 ppm and (b) U ~1.7 ppm and Th ~25.4 ppm. Surface

http://dx.doi.org/10.5772/61963

#### 3.2.2. *Perth metropolitan area*

The concept of recovering geothermal energy from the Yilgarn Craton is significantly enhanced by the presence of the city of Perth, the capital city of Western Australia, located near the western margin of the craton. The Darling Range is a physiographic expression of the Yilgarn Craton on the eastern edge of the Perth metropolitan area. Before 2010, only a few values of heat generation were available in the literature for the Yilgarn Craton [4,16]. However, studies have been carried out by the Western Australian Geothermal Centre of Excellence (WAGCoE) from 2009 to 2013 [42]. This centre carried out extensive studies on the potential of geothermal energy in Western Australia, and perhaps culminated in the concept of the "geothermal city" by Regenauer-Lieb et al. [43] and geothermal modeling of the Perth Basin by Schilling et al. [44].

The interest in geothermal energy was initially directed toward electricity generation. However, Ballesteros [45] noted that there had been steady growth in the direct use of geothermal heat in Western Australia, especially in the Perth metropolitan area. He reported that 12 sites were using, or were in development, geothermal heat for community or industrial projects. Pujol et al. [46] and Ricard and Pujol [47] have reviewed the previous 20 years of exploitation of the Yarragadee aquifer in the Perth Basin for direct-use geothermal heat. These authors have indicated how the face of geothermal energy in Western Australia has changed from a perception of electricity creation to direct heat energy extraction.

### 3.3. Northern Perth basin

The northern Perth Basin is bounded by the Northampton Complex to the north and the Yilgarn Craton to the east. Reasonable U and Th data are known from the Northampton Complex and airborne data are available for the Yilgarn Craton to the east. The Northampton Complex (see Figure 1 for location), which is part of the Pinjarra Orogen, has been mapped on the Geological Survey of Western Australia's Geraldton and Ajana 1:250,000 geological sheets [48]. The Northampton Complex forms the pre-sedimentary "basement" of the northern part of the Perth Basin, which is considered to possess appreciable geothermal energy potential, due to high temperatures observed within the sediments of the adjacent basin [12,50]. The high temperatures are presumed to be caused by elevated heat generation in the underlying Northampton Complex rocks.

Little recent geological work has been carried on the Northampton Complex. The Northampton Complex (see Figures 1 and 2 for location) has been mapped on the Geraldton and Ajana 1:250,000 geological map sheets [48], and is part of the larger Pinjarra Orogen, which lies along the western margin of onshore Western Australia. On the Ajana 1:250,000 geological sheet, the Northampton Complex is dominated by rocks described as granulite and gneiss. These rocks have been metamorphosed under pressures in the range of 600–900 MPa (6–9 kBar) and temperatures in the range of 600– 800 °C (upper amphibolite-to-granulite facies metamorphic conditions; Sanders and McGuiness [50]). These rocks are not dissimilar to those of the Leeuwin Complex, which is also considered to be part of the Pinjarra Orogen. The merged thorium aero-radiometric map (Figure 2) suggests much of the Northampton Complex region, which consists of rocks with greater than 19.5 ppm thorium, noting that the proposed upper cutoff for the aero-radiometric maps for thorium is 19.5 ppm, as discussed above. However, also mentioned above and evident from ground-based studies in the Leeuwin Complex and Darling Range, one should be aware that 19.5 ppm appears to be the limit on the calibration technique of aero-radiometric observations for thorium, rather than a peak value that might be observed at the earth's surface. Laboratory-based values of concentration reported by Sanders and McGuiness [50] suggest that some localities may be significantly higher (i.e., > 39 ppm) than the aero-radiometric threshold maximum of 19.5 ppm.

Sanders and McGuiness [50] reviewed the geochemistry of the Ajana 1:250,000 geological sheet, and therein presented the valuable data concerning the surface distribution of thorium in the regolith. For the purposes of this study, the regolith is defined as unconsolidated or indurated weathered rock, and includes residual and transported material that can cover and obscure the underlying bedrock. The thorium concentrations reported in this review are from the Geological Survey of Western Australia map of Sanders and McGuiness [50], and the values were measured by the recognized laboratory methods referenced in the Sanders and McGuiness' study. Their map is reproduced as Figure 15 in this chapter. The study presented a systematic grid of surface thorium concentrations in outcropping rocks at a spacing of approximately 5–10 km, nominally one sample per 16 km$^2$. The map of Sanders and McGuiness displays a range of thorium concentration between approximately 5 and > 39 ppm. Unfortunately, the upper limit of thorium concentration in the study, although acknowledged to be greater than 39 ppm, is not reported in their report.

Rocks from the Northampton Complex are very likely to extend under the north Perth Basin, and they probably contribute to the high temperatures observed in petroleum wells in that region [12,35]. Little definitive work has been carried out to date on the influence of radiogenic elements in the basement rocks upon elevated temperatures in the sediments of the north Perth Basin. Basement rocks have also been intersected in a number of petroleum wells. Table 4 shows an analysis of the gamma-ray log response found in the basement rocks in this part of the basin. The table also shows the gamma-ray log response for high-organic shale formations and some sandstones in the observation well for comparison with the basement rocks. Equation 8 is used to convert gamma-ray (GR) data from the wells to heat generation ($A_0$) data. Heat generation in basement rocks determined by this method is seen to be in the range of 2.0–5.6 $\mu Wm^{-3}$.

Thorium values mapped on the Ajana 1:250,000 geological sheet from Sanders and McGuiness [50] reflect relative heat generation as expressed in Eq. (4). Surprisingly, high Th is also found in the sedimentary rocks on the margin of the northern Perth Basin, but it does not appear to be especially dominant in the surrounding igneous and metamorphic rocks. The provenance of high radiogenic elements in sediments of the northern Perth Basin (perhaps also the southern Carnarvon Basin) is currently an imponderable. Further studies are needed to fully understand the geology, geochemistry, and geophysics of the northern Perth Basin and Northampton districts.

| Well Name | Basement Depth (m) | Structural Location | GR av. (API) | GR max. (API) | Range of Heat Generation, Ao ($\mu Wm^{-3}$) | Reference lithology GR av. (API units) |
|---|---|---|---|---|---|---|
| Arramall 1 | 2225 | | 280 | 280 | **4.5** | 260, PCS |
| Arrowsmith 1 | 3420 | | 180 | 230 | **2.9 – 3.7** | 120, HS; 100, K |
| Bookara 1 | 852 | | 150 | 195 | **2.4 – 3.1** | 135, K |
| Bookara 3 | 1560 | | 168 | 204 | **2.7 – 3.3** | 144, K |
| Sue 1 | 3054 | | 255 | 265 | **4.1 – 4.2** | 270, PCS; 140, V |
| Conder 1 hot sandst. | 203-213 | | > 400 | > 400 | **> 6.4** | 140, K |
| Conder 1 basement | 229 | | 180 | 230 | **2.9 – 3.7** | 140 K |
| Jurien 1 | 3280 | | 140 | 165 | **2.2 – 2.6** | 150, PCS |
| Wendy 1 | 1211 | | 220 | 280 | **3.5 – 4.5** | 40, TS |
| Cadda 1 | 2743 | | 125 | 200 | **2.0 – 3.2** | 150, PCS |
| Woolmulla 1 | 2804 | | 50 | 160 | **0.8 – 2.6** | 70, PCS |
| Beharra 1 | 2075 | | 155 | 162 | **2.5 – 2.6** | 140 PCS |
| Drakea 1 | 3048 | | 225 | 255 | **3.6 – 4.1** | 195, K; |
| Cliff Head 1 | 1480 | | 170 | 195 | **2.7 – 3.1** | 180, PCS |
| Connolly 1 | 450 | | 195 | 218 | **3.1 – 3.5** | 180, K; 200, PCS |
| Wattle Grove 1 | 799 | | 260 | 350 | **4.2 – 5.6** | 150 K |
| Mountain Bridge 1 | 3385 | | 250 | ? 350 | **4.0 - ? 5.6** | 150 K |
| North Yardarino 1 | 2190 | | - | - | **-** | NPD |
| Woodada 19 | 2035 | | - | - | **-** | NPD |
| Dunnart 1 | ? | | ? | ? | **?** | Dubious data |
| Hovea 2 | ? | | ? | ? | **?** | Data accessibility ? |
| Dongara 6 | 1540 | | - | - | **-** | NPD |

**Reference lithology code for Table 4.** HS: Holmwood Shale (Permian); PCS: Permian coals and shale; IRCM: Irwin River Coal Measures; V: Volcanic intrusives (? Cretaceous); K: Kockatea Shale (Triassic); TS: Tumblagooda Sandstone; NCR: No Petrophysical Data (for gamma-ray). See Mory and Iasky (1996) for reference to stratigraphic units.

**Table 4.** Northern Perth Basin with U, Th, K & $A_o$ derived from petroleum well data. The figure shows showing gamma-ray log (GR) responses from basement rocks with established crystalline basement intersections and derived sediments. Reference gamma-ray units (API) within the overlying formations for various wells in the Perth Basin are shown for confidence in gamma-ray logging in API units. The heat generation calculation, derived from observed log-derived GR data is $A_o$ = 0.015 (GR) (see text for explanation).

## 4. Pilbara region

The Pilbara region of Western Australia contains some of the state's most significant mineral deposits. However, difficulties can arise for the development of these mineral resources due to the remoteness and lack of economic energy supply. The energy supply for mines at remote locations in this region is commonly sourced by diesel-powered generators, although some mines are supplied with gas from the North West Shelf. Geothermal energy may provide an economic alternative, or additional energy supply, if access to deep radiogenic granitoids is available. Hot Dry Rock geothermal systems are one of three ways that are considered viable for using geothermal energy in Australia. The three systems are (1) direct heat extraction, (2) Hot Sedimentary Aquifers (HSA) energy conversion, and (3) the Hot Dry Rock geothermal regime [12,51]. The HSA system, as well as direct heat extraction, is currently being explored by a number of companies as a possible system for both power generation and air conditioning in Western Australia. An HDR system is being exploited in the Cooper Basin by the geothermal company Geodynamics Limited. The HDR system entails injecting cool water into hot fractured rocks at depths of about 4 km, allowing the water to percolate through the hot rocks (which heats the water), and then extracting the hot water for geothermal energy applications. Geothermal explorers commonly seek temperatures in excess of 200 ºC for such systems. The hot water is fed into various types of turbines for electricity generation; the efficiency of electricity generation is dependent on the temperature of the fluid being introduced into the turbine.

Two regions of high radiogenic heat generation can be seen in Figure 2. The two hot regions are identified to be within the Numbana and Coolegong mesogranite intrusions. These intrusions are late-stage, highly fractionated granites [52,53] known to possess high U and Th concentrations, which crop out at the surface. Table 1 shows U and Th versus radiogenic heat generation ($A_0$) of samples of the Numbana and Coolegong granites (data courtesy of the Geological Survey of Western Australia). An excellent linear trend has been found between U and heat generation ($A_0$). The linear regression curve through U and $A_0$ data from Pilbara and Yilgarn Craton granitoids exhibit a simple relationship [16]:

$$U = 2.2A_0 - 5.4, \tag{9}$$

where $U$ is the uranium concentration (ppm) and $A_0$ is the radiogenic heat generation (μWm$^{-3}$); the coefficient of linear regression ($R^2$) is 0.92. As mentioned previously, the U and Th (essentially reflecting heat generation) of the rock samples are generally higher than the values observed from the airborne data [16]. These data suggest that the radiogenic heat generation in the mesogranites of the Pilbara Craton may be as high as 13 μWm$^{-3}$, which is in the vicinity of those observed in the Leeuwin Complex in south west part of Western Australia.

Figure 18 shows temperature models displaying expected temperature versus depth within the two granitoids in the Pilbara region. The numbers on each curve represent the thickness (in km) of a granitoid with a uniform heat generation of 12 μW m$^{-3}$. Temperatures at depths within the granitoids of 3–4 km may be over 100 ºC, but these require drilling with large-diameter bore holes to those depths through crystalline rocks. Currently, this does not present an economic development.

New generations of turbines, for example, Kalina cycle and variable phase turbines [54,55], are allowing the use of lower-temperature geofluids for electricity generation. A stand-alone variable phase turbine has been designed to generate about 1 MW of electricity with input fluid temperatures of about 110 ºC, although greater generation efficiencies are obtained at higher temperatures. These new generation of turbines may hold significant implications for generating power at Pilbara mining sites. The key ingredient is the proximity of a hot granite to a mine location. Figure 2 shows the approximate radiogenic heat generation of surface rocks on the Marble Bar 1:250,000 map sheet. The map is generated from calibrated uranium, thorium, and potassium elemental abundances determined from airborne radiometric surveys [9,34]. The U, Th, and K abundances are converted to approximate radiogenic heat generation by a method described by Kappelmeyer and Haenel [23].

## 5. Carnarvon and Canning basins

Apart from Perth Basin, HAS geothermal resources have been shown potentially to exist in the Carnarvon and Canning basins [12,51]. Some further work has been carried out for this review on the Carnarvon Basin, which explores, in a preliminary sense, the anomalous elevated temperature regime in the southern Carnarvon Basin. For the Canning Basin (Figure 1), little new geothermal insight has been gained since the study of Hot Dry Rocks [57], which was commissioned by the Geological Survey of Western Australia. Beardsmore and Cull [7] have reported some important heat generation data derived from gamma-ray logs in that basin. Their approach has been used earlier in this chapter to attempt to understand heat generation and the presence of radiogenic elements. Given the lack of new data from the Canning Basin, only the Carnarvon Basin will be discussed further.

### 5.1. Carnarvon basin

High temperatures have been recognized in the Southern Carnarvon Basin (Figure 1). These elevated temperatures have been reported in various studies [12,51,56]. This region is probably the most prospective region in Western Australia for geothermal energy, behind the Vasse Shelf region in the southern Perth Basin. However, very little geothermal exploration has been carried out in the region. However, substantial evidence supports the premise that the Carnarvon Basin may be able to supply the Pilbara mining communities with not only (1) long-term geothermal-sourced electricity (looking beyond North West Shelf-sourced gas supply, i.e., beyond perhaps 2050), but also (2) intermediate-term nearby resource projects and tourism enterprises. It is of note that Davidson [58,59] described the Peninsular Hot Spring tourist enterprise in southern Victoria, which entertains an estimated $75 million per annum regional economic benefit. Such enterprises are potentially available to environmentally sensitive, but tourist intensive, coastal resort sites from Carnarvon north to Exmouth.

### *5.1.1. Geology and geophysics*

The geology of the southern Carnarvon Basin is well understood, having been described by Hocking et al. [60], Hocking [61], and the geothermal potential by Ghori [12]. The geology of the North West Cape, which is especially relevant to the current article, was studied by Malcolm et al. [62]. It has been shown from petroleum drilling data that an unusually hot region exists along the coastal region from Shark Bay to the Exmouth Peninsula, and maybe for some distance inland toward, and perhaps including, the outcropping Pilbara Craton. It is reasonable to expect that the Pilbara Craton underlies a reasonable amount of the onshore Carnarvon Basin between Shark Bay and Exmouth, including the North West Cape. Further, it is expected that the Pilbara Craton may contain some radiogenic granitoids in this region. There appears to be no detailed scientific study of why the elevated heat flow occurs in this region, despite it being covered since 2010 by seven GEPs. According to the studies by Ghori [12,51], temperatures in excess of 150 ºC may exist between drillable depths of 3000–4000 m, and these wells have a geothermal gradient in the range of 60–100 ºC/km (figure 2 from Ghori [12]).

### *5.1.2. A geothermal puzzle*

The unanswered geothermal puzzle of the southern Carnarvon Basin is centered on the cause of the elevated geothermal gradients in the Exmouth and Gascoyne subbasins. Essentially, the puzzle may condense into whether the elevated temperatures reported by Ghori [12,51] are caused by underlying radiogenic granitoids or by the flow of deep hot groundwater into surface sediments. No extensive studies have been carried out to understand this phenomenon. The source of the observed high heat flow in petroleum wells appears to be generated from high radiogenic granitoids that underlie the deep sedimentary blanket in the region. However, this is only a circumstantial evidence, and without sufficient further investigation, it cannot be stated with certainty. Convective heat transport should not be ruled out as the cause of these elevated temperatures. Comparison to the surface geological units to elsewhere in Western Australia suggests that the sedimentary cover may have a surface heat generation in the vicinity of 1 $\mu Wm^{-3}$ (see Table 4, and Beardsmore and Cull [7]). On the adjacent Pilbara Craton, outcropping granitoids possess heat generation values of up to 5 $\mu Wm^{-3}$ in the adjacent Pilbara Craton; these may translate up to 12 $\mu Wm^{-3}$ in specific geologic units on the ground. Therefore, the granitoids that are interpreted to underlie the Carnarvon Basin in the cross section in Figure 19 may possess a heat generation of up to 12 $\mu Wm^{-3}$ or more. This heat generation value is typical of radiogenic granitoids underlying the Perth Basin, and also the central Australian Cooper Basin [6], where geothermal energy is currently being developed. The WAPET Cape Range petroleum exploration wells are known to encounter bottom-hole temperatures in excess of 140–160 ºC at depths between 4300 and 4600 m; these temperatures are probably greater after fluid stabilization in the drill hole. Figure 19 shows that the total sedimentary blanket over interpreted high heat-generating granites may be up to 9 km [61,62].

### *5.1.3. Temperature modeling*

Based on the cross section [56] shown in Figure 19, several numerical models of the temperature versus depth for various heat-generation scenarios are shown in Table 5a. These are based on the previously described 1D models for such situations as described in Eqs (1) and (2). The assumptions used to generate Table 5a are shown in Table 5b, and they are based on the surface radiogenic heat data derived from Figure 2 and Eq. (4), and also studies reported elsewhere in the literature [4,5,16,61]. Ghori [12] has shown from petroleum wells that the temperature gradient in the southern Carnarvon Basin can range between 30 and 100 ºC/km, and these are consistent with the temperatures generated in the models in Table 1. If convection is not involved, then the governing factor appears to be the heat generation of underlying granites. However, a disturbing observation from hydrogeological studies is that there are significant indications of hot artesian water movement in this basin. The static models in Table 5a demonstrate that temperatures at depths greater than 3000 m may have temperatures significantly greater than 200 ºC, depending on the heat generation of underlying granite bodies. In the upper side of the models, one is looking at full steam-type turbines for the generation of electricity, which are the current optimal geothermal energy-producing technology, and is common place in New Zealand, Indonesia, and California. A caveat needs to be placed on the geothermal activity in this basin centered around the balance of radiogenic heat input versus convective heat input. More investigation is required.

| **Heat Generation ($\mu$W m$^{-3}$)** | **Geothermal Gradient (°C km$^{-1}$)** | **Temp @ 2 km (°C)** | **Temp @ 3 km (°C)** | **Temp @ 4 km (°C)** |
|---|---|---|---|---|
| 3 | 42 | 104 | 146 | 188 |
| 6 | 60 | 140 | 200 | 260 |
| 9 | 78 | 176 | 256 | 332 |
| 12 | 96 | 212 | 308 | 404 |
| | | (a) | | |

| **PARAMETER** | **VALUE (units)** | **COMMENT** |
|---|---|---|
| Sediment thickness | 4000 (m) | Varies between 2 and 8 km |
| "Hot" granitoid thickness | 6000 (m) | May vary between 2 and 8 km |
| Basal heat flow | 24 ($\mu$Wm$^{-3}$) | Assumed from Perth Basin |
| Surface temperature | 20 (°C) | Can be higher |
| Thermal conductivity | 3 (Wm$^{-1}$°C$^{-1}$) | Variable between 2 and 5 units |
| | (b) | |

**Table 5.** Carnarvon Basin modeling results (a) temperatures and (b) assumptions. See also the geological cross-section in Figure 19 for geological consideration discussed in the text.

## 6. Time-variant considerations

It may be worth examining some time-variant models for heat flow and temperature variation for Western Australian geothermal scenarios. These are largely based on applications of Eq.

(3) [24], which considers the case where a layer of rock with high heat generation is rapidly placed at the earth's surface at a time $t = 0$. A typical situation for Western Australia may occur where lateritic enrichment occurs in the Tertiary Period. The question being posed is: does this enrichment in U and Th alter the temperature profile in the vicinity of the chemical alteration process, and, if so, how great is the temperature disturbance and how long does it take to stabilize? An initial intuitive guess would be that stabilization occurs fairly rapidly and there is not much disturbance to the temperature or heat flow regime. A lateritic profile in the Darling Range may be up to several hundred meters thick [40], but generally less than 100 m. A weathering profile on the Leeuwin Complex may be of similar thickness [63]. The question posed revolves around the difference that occurs for a heat-generating profile of thickness between 50 and 300 m at the surface. We can use the Meelup Park enrichment value (Table 1) of 40 μW m$^{-3}$ as an initial test.

Carslaw and Jaeger [24] propose that the surface geothermal gradient for a geometry shown in Figure 3, where a heat-generating layer (injection of fluids or erosional enrichment) occurs rapidly, can be described by an equation of the type:

$$\Delta Q_R = 2A_0(\alpha t)^{\frac{1}{2}}\left\{1/(\pi)^{\frac{1}{2}} - \operatorname{ierfc}\left[L/(4\alpha t)^{\frac{1}{2}}\right]\right\}, \tag{10}$$

where $\Delta Q_R$ is the heat flow contribution of the radiogenic layer in units of mW m$^{-2}$, $A_0$ is the uniform heat generation in the layer in μW/m$^3$, $\alpha$ is the thermal diffusivity of the rocks in m$^2$s$^{-1}$, $t$ is the time in seconds, $L$ is the thickness of the radiogenic enrichment layer in m, and ierfc[$x$] is the integrated error function. This equation has been modified from the Carslaw and Jaeger [24] equation 6 by multiplication by thermal conductivity ($K$) to be expressed as heat flow rather than thermal gradient. Carslaw and Jaeger also indicate that this equation will reduce to the simple form of $\Delta Q_R = A_0 L$, where time becomes very large (> 10 million years) and $L$ is less than 50 km. This was referred to earlier with regard to Eq. (1).

Some simple calculations with Eq. (9) indicate that stabilization of an enrichment layer, such as observed at Meelup Park (see Table 1 and Figure 9 for magnitude of Th anomaly) in the Leeuwin Complex and the Vasse Shelf, southern Perth Basin, will occur well within 1 million years of enrichment, where a 100 m of enrichment zone is assumed. A zone of 100 m may be expected, based on geological and environmental mapping [63]. Similar durations for thermal stabilization may also be expected for the lateritic zones in the Darling Range in the central Perth Basin, although such high levels of enrichment are not observed from current observations (see Table 1). These conclusions, based on numerical modeling, are consistent with expectations. In the case where $\Delta Q_R = A_0 L$, which is the stabilized case above, it is interesting to understand the influence that an enriched zone may have on heat flow. Table 6 shows a range of incremental heat flow ($\Delta Q_R$) values that will occur for a range of heat generation ($A_0$) and enrichment layer thicknesses ($L$). The table indicates that enrichment zones may contribute an appreciable input to localized heat flow, especially if the thickness is greater than 100 m and heat generation is greater than 80 μW m$^{-3}$.

http://dx.doi.org/10.5772/61963

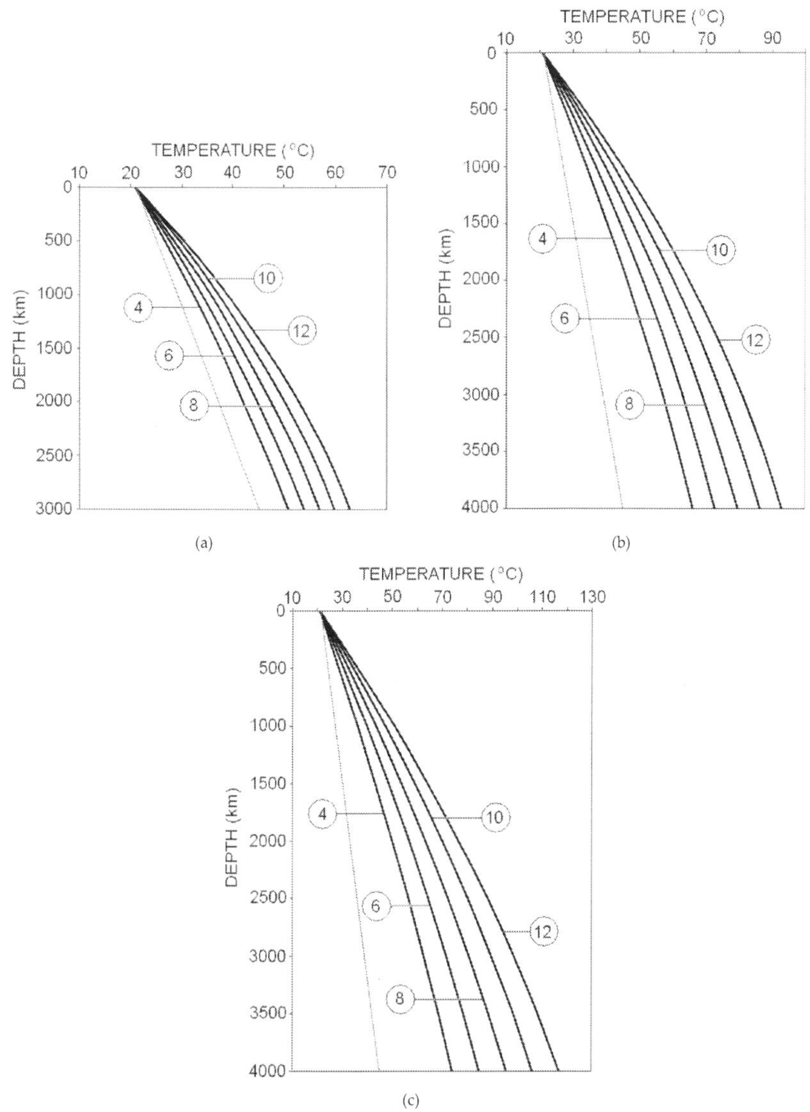


**Figure 14.** Temperature versus depth for the Darling Range [16]. The values shown in the circles are the assumed uniform heat generation in the radiogenic (granitoid) layer in units of $\mu W\ m^{-3}$. Figure 14(a) shows temperature versus depth with the radiogenic layer of depth extent of 3 km. Figure 14(b) shows temperature versus depth with the radiogenic layer of depth extent of 4.5 km. Figure 14(c) shows temperature versus depth with the radiogenic layer of depth extent of 6 km. The curves in red on 14a, 14b, and 14c represent the temperature versus depth profile, where there is no heat generation in the granitoid layer.

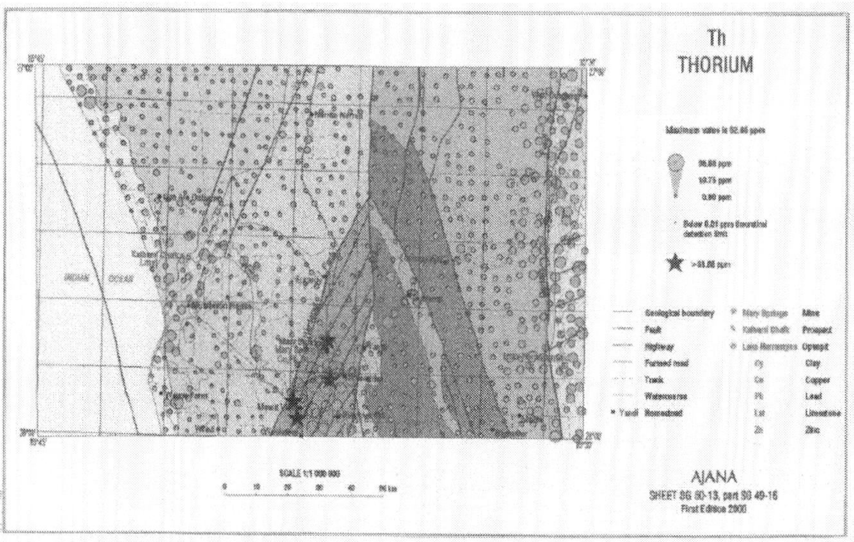


**Figure 15.** Thorium values Ajana 1:250,000 geological sheet from Sanders and McGuiness [50] are shown in this figure. The map also reflects the main contributing sources to heat generation in the vicinity of the northern Perth Basin and the rocks expected to underlie it. High Th is found in the sedimentary rocks on the margin of the northern Perth Basin, and it does not appear to be significantly less than the surrounding igneous and metamorphic rocks. The provenance of high radiogenic elements in sediments of the northern Perth Basin (perhaps also the southern Carnarvon Basin) is currently an imponderable. More work is definitely needed to understand the geology, geochemistry, and geophysics of the northern Perth Basin and Northampton district

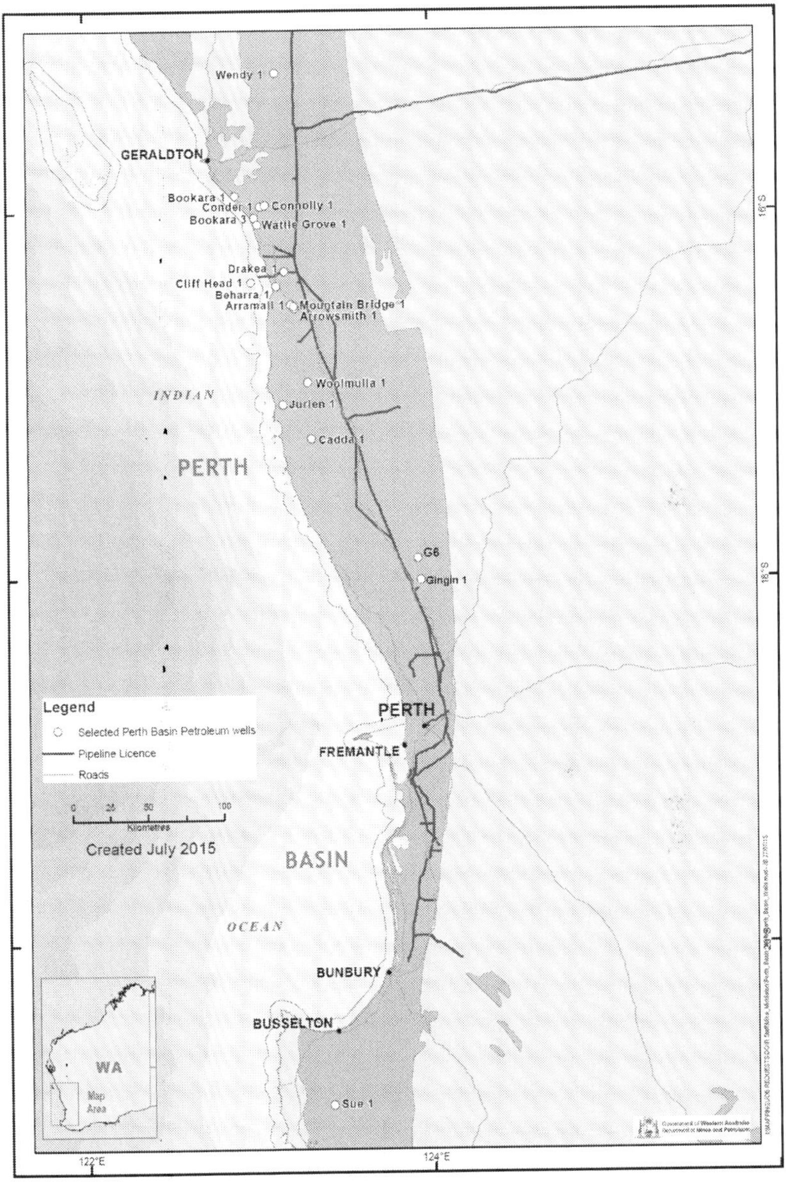


**Figure 16.** North Perth Basin well locations with basement intersections for potential heat generation estimations (combined with Table 4).

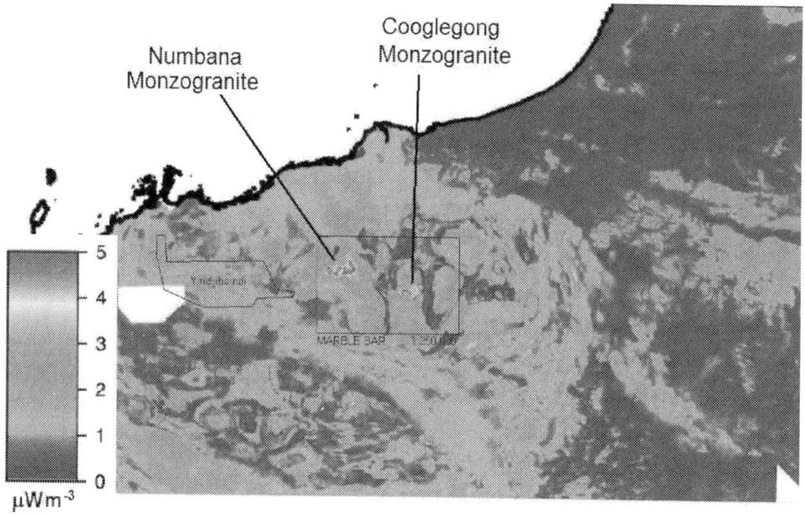


**Figure 17.** Pilbara Region hot granites. A map showing the integration of airborne U, Th, and K concentrations to reflect surface heat generation in the Pilbara region) [34]. The "hot" Numbana and Coolegong granitoid bodies are seen as the areas as shown in the image.

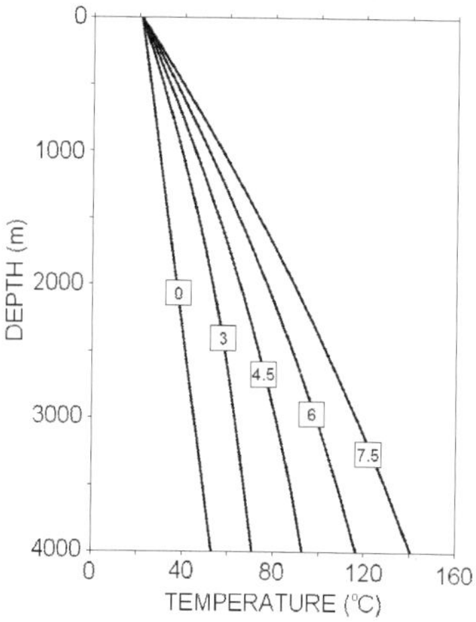


**Figure 18.** Pilbara temperatures for hot granites. The curves represent temperature versus depth for a radiogenic granitoid of uniform heat generation of 12 $\mu W\ m^{-3}$. The numbers on the curves represent the thickness of the radiogenic (granitoid) layer.

[22] Minty, B., Franklin, R., Milligan, P., Richardson, L.M. and Wilford, J., 2009, The radiometric map of Australia, *Exploration Geophysics*, 40, 325–333.

[23] Kappelmeyer, O. and Haenel, R., 1974, Geothermics with Special Reference to Application, Bebrüder Borntraeger.

[24] Carslaw, H.S. and Jaeger, J.C., 1959, Conduction of Heat in Solids, Oxford University Press.

[25] Jessop, A.M., 1990, Thermal Geophysics, Developments in Solid Earth Geophysics, Elsevier, Amsterdam, 306p.

[26] Ryback, L., 1986, Amount and significance of radioactive heat sources in sediments. In: Thermal Modelling in Sedimentary Basins, Collections Colloques et Seminaires (J. Burrus, Ed.) 44, Editions Technip, Paris.

[27] Bücker, C. and Ryback, L., 1996, A simple method to determine heat production from gamma-ray logs, *Marine and Petroleum Geology*, 13, 373–375.

[28] SPE, 2015, http://petrowiki.org/Gamma_ray_logs.

[29] Crostella, A., and Backhouse J., 2000, Geology and petroleum exploration of the central and southern Perth Basin, Western Australia: *Western Australia Geological Survey, Report 57*, 85p.

[30] Cockbain, A.E., Hocking, R.M. and Myers, J.S., 1990, Geological Evolution, Phanerozoic, in *Geology and Mineral Resources of Western Australia, Western Australia Geological Survey*, Memoir 3, 735–755.

[31] Middleton, M.F., Wilde, S.A., Evans, B.J., Long, A., Dentith, M. and Morawa, 1995, Deep seismic reflection traverse over the Darling Fault Zone, Western Australia, *Australian Journal of Earth Science*, 42, 83–93.

[32] Iasky, R. P., 1993, A structural study of the southern Perth Basin, Western Australia: *Western Australia Geological Survey, Report 31.*

[33] Mory, A.J. and Iasky, R.P., 1996, Stratigraphy and structure of the onshore Perth Basin, Western Australia: *Western Australia Geological Survey, Report 46.*

[34] Middleton, M.F., 2011, Temperatures within outcropping, radiogenic granites: Implications for geothermal energy in Western Australia, in: Middleton, M. and Gessner, K. (Eds) *West Australian Geothermal Energy Symposium Abstracts*, 21 & 22 March 2011, v. 1, 20 (ISBN 978-1-74052-230-41.

[35] Hot Dry Rocks, 2008, Geothermal energy potential in selected areas of Western Australia (Perth Basin), a report prepared for the Department of Industry and Resources, Western Australia. *Geological Survey of Western Australia, Statutory Petroleum Exploration Report*, G31888 A2.

[36] Schön, J.H., 1996, Physical Properties of Rocks: Fundamentals and Principles of Petrophysics, Pergamon.

[37] Wharton, P.H., 1982, The geology and hydrogeology of the Quindalup boreline, southern Perth Basin, Western Australia, *Geological Survey of Western Australia*, Record 1982/2.

[38] Playford, P. E., Cockbain, A. E. and Low, G. H., 1976, Geology of the Perth Basin, Western Australia: *Western Australia Geological Survey, Bulletin* 124.

[39] Wilde, S.A., Middleton, M.F. and Evans, B.J., 2006, Terrane accretion in the southwestern Yilgarn Craton: evidence from a deep seismic crustal profile, *Precambrian Research*, 78, 179–196.

[40] Annand, R.R. and Butt, C.R.M., 2003, Distribution and evolution of 'laterites' and lateritic weathering profiles, Darling Range, Western Australia, *Australian Geomechanics*, v. 36, No. 4, 41–58.

[41] Middleton, M.F., 2010, The Gillingarra temperature anomaly, northern Perth Basin, Australia, *Expanded Abstracts, Proceedings of SEG International Exposition and 80th Annual Meeting, Denver, Colorado*, October 2010.

[42] Regenauer-Lieb, K. and the Team of the Western Australian Centre of Excellence, 2011, Geothermal Cities, Australian Geothermal Energy Conference 2011, Geoscience Australia Record 2011/43, (A. Budd, Ed.), 185–188.

[43] Regenauer-Lieb, K. and the Team of the Western Australian Centre of Excellence, 2012, The Western Australian Geothermal Research Centre and its future, in: Middleton, M. and Gessner, K. (Eds) *West Australian Geothermal Energy Symposium Abstracts*, 2 & 3 April 2012, v. 2, 23–24 (ISBN 978-1-74052-250-2.

[44] Schilling, O, Sheldon, H.A., Reid, L.B. and Corbel, S., 2013, Hydrothermal models of the Perth metropolitan area, Western Australia: implications for geothermal energy, *Hydrogeology Journal* 21 (3), 605–621.

[45] Ballesteros, M., 2012, personal communication, concerning geothermal developments in the Perth metropolitan area.

[46] Pujol, M., Ricard, L., and Bolton, G., 2015, 20 years of exploitation of the Yarragadee aquifer in the Perth Basin of Western Australia for direct-use geothermal heat, *Geothermics*, v. 57, 39–55.

[47] Ricard, L and Pujol, M., 2015, Geothermal production in the Perth Metropolitan Area, 20 years of success. Petroleum in Western Australia, September issue, 2015.

[48] Hocking, R. M., van de Graaff., W. J. E., Blockley, J. G. and Butcher, B.P., 1982. Ajana, W.A.: *Western Australia Geological Survey*, 1:250 000 Geological Series Explanatory Notes.

[49] Ghori, K.A.R. and Gibson, H., 2011, Geothermal simulation, Perth Basin, Western Australia, in: 2011, *Proceedings of the 2011 Australian Geothermal Energy Conference*, 16–

18 November, Melbourne, Geoscience Australia, Record 2011/43, (A. Budd, Ed.), 77–80.

[50] Sanders, A. J. and McGuiness, S. A., 2000, Geochemical mapping of the Ajana 1:250 000 map sheet: *Western Australia Geological Survey*, 1:250 000 Regolith Geochemistry Series Explanatory Notes, 55p.

[51] Ghori, K.A.R., 2008, Perth Basin's geothermal resources, *Australian Geothermal Energy Conference 2008, Extended Abstracts*, 55–61.

[52] Van Krandendonk, M.J., Smithies, R. H., Hickman, A.H. and Champion, D.C., 2007, Review: secular tectonic evolution of Archean continental crust: interplay between horizontal and vertical processes in the formation of the Pilbara Craton, Australia, *Terra Nova*, 19, 1–28.

[53] Van Krandendonk, M.J., 2011, Personal communication concerning the geochemical database of the Geological Survey of Western Australia.

[54] Welch, P. and Boyle, P., 2009, New turbines to enable efficient geothermal power plants, *Geothermal Research Council (GRC) Transactions*, 33, 765–772.

[55] Welch, P., Boyle, P., Sells, M. and Murillo, I., 2010, Performance of new turbines for geothermal power plants, *Geothermal Research Council (GRC) Transactions*, 34, 1091–1096.

[56] Middleton, M.F., 2013, Implications of high heat flow and temperatures for geothermal energy in the southern Carnarvon Basin, Petroleum in Western Australia, April issue 2013. Hot Dry Rocks, 2009, Geothermal Energy Potential in Selected Areas of Western Australia (Canning Basin), a report prepared for the Department of Mines and Petroleum, *Western Australia Report* DOIR0681008.

[57] Davidson. C., Anderson, T. and Stanley, D.R, 2011, Development of Geothermal Waters for Recreational Purposes – Mornington Peninsula Australia, Australian Geothermal Energy Conference 2011, Geoscience Australia Record 2011/43, (A. Budd, Ed.), 49–52.

[58] Davidson, C., 2012, Geothermal beyond power – Opportunities for geothermal resource development, West Australian Geothermal Energy Symposium, Abstracts 2012, 2–3 April 2012, (M. Middleton and K. Gessner, Eds), ISBN 978-1-74052-250-2, p. 4.

[59] Hocking, R. M., Moors, H. T. and van de Graaff. W. J. E., 1987, Geology of the Carnarvon Basin, Western Australia: *Western Australia Geological Survey*, Bulletin 133.

[60] Hocking, R., 1990, The Carnarvon Basin, in Geology and Mineral Resources of Western Australia, *Western Australia Geological Survey*, Memoir 3, 457–495.

[61] Malcolm, R.J., Pott, M.C. and Delfos, E., 1991, A new tectono-stratigraphic synthesis of the North West Cape area, *APEA Journal*, 31, 154–176.

[62] Lennard, E.L., 1991, YALLINGUP SHEET 1930 IV and part Sheet 18301, Geological Survey of Western Australia. 1:50 000 Environmental Geology Series.

Chapter 4

# Analysis of Geochemical and Production Well Monitoring Data — A Tool to Study the Response of Geothermal Reservoirs to Exploitation

Rosa María Barragán, Víctor Manuel Arellano and David Nieva

Additional information is available at the end of the chapter

http://dx.doi.org/10.5772/61586

**Abstract**

Geothermal resources used to generate power are complex systems. The hydrothermal, high-temperature geothermal reservoirs, which are the only commercially exploited ones up to now, are usually found several (1-3 km) kilometers deep, and they should be exploited in a sustainable way. Thus, in order to support decisions on optimal exploitation policies, efforts focused on investigating how reservoirs respond to exploitation are routinely made. The better strategies always involve a compromise of controlling energy extraction from geothermal reservoirs without overexploiting the resource. Currently, methods based on the analysis of monitoring data of either the geochemical characteristics of fluids discharged (water and steam) or production data gathered from wells have been used to assess the reservoir performance, making it possible to predict the occurrence of negative processes in terms of production. However, by the analysis of combined geochemical and production data through simulation of wells, the well-bottom thermodynamic conditions of fluids are included in the study, allowing more reliable results to be obtained. Besides, the comparison of actual patterns of behavior of chemical and production indicators with those characteristic for typical processes, helps in identifying different physical phenomena and in deciding which is dominant in the case of the occurrence of more than one processes.

**Keywords:** Geothermal fields, reservoir exploitation, thermodynamic conditions, fluid geochemistry, well simulation

# 1. Introduction

Geothermal resources should be exploited in a sustainable way; therefore, geochemical studies focused on investigating how reservoirs respond to exploitation are routinely performed in order to support decisions on optimal exploitation policies [1]. As high-temperature geothermal reservoirs are usually found several (1-3 km) kilometers deep, indirect methods based on the monitoring of both: the geochemical characteristics of fluids discharged and production data of wells are used to assess their performance. The chemical and isotopic compositions of fluids produced by wells at an early exploitation stage are useful to develop preliminary geochemical models for the reservoirs from which it is possible to obtain reference values of geochemical indicators [2]. A detailed reservoir conceptual model often includes information on well production, which is obtained when wells are evaluated in order to characterize the resource in terms of electricity generation capacity. This task includes the study of well production data under different operating conditions during the development stage of the field to obtain thermodynamic reservoir parameters such as pressure, enthalpy, and steam quality among others, which constitute important reference values [3]. During exploitation, the variations in well production and geochemical data with respect to such references can be correlated to typical patterns of behavior indicative of the occurrence of physical reservoir processes.. In this work, both geochemistry of fluids and well production data are combined to investigate the response of geothermal reservoirs to exploitation. As usually either geochemical or production data are used, this methodology combining both disciplines is documented and illustrative examples for Mexican fields are described [4-6].

# 2. Methodology

## 2.1. Well production over time

The economic aspects of a geothermal project are strongly influenced by the well production since the heat (which is the main objective to be exploited in geothermal fields) is transported from the reservoir to the wellhead by the fluids. The production of liquid and steam from geothermal wells usually changes over time due to exploitation; therefore, the analysis of well production behavior over time provides valuable information to identify reservoir physical processes. According to [7] the "normal" behavior of the mass flow of a well at constant wellhead pressure or constant throttle should decline smoothly with time, as reservoir pressure falls. Abnormal behavior appears as a change in mass flow and this could be a sign of mechanical damage in the well such as casing break, liner collapse, or scaling. For this reason, the rates of decline or increment (even temporal) in well production of at least the most representative wells of the field should be assessed.

In order to study and make use of well production data to estimate and predict production decline rates in geothermal wells, two methods have been used. (a) The decline curves method [8], which has been successfully used in the oil industry, has also been used in geothermal fields although with limited success. For example in The Geysers, California (USA), the decline curves method allowed the electricity capacity of a specific area to be estimated in order to optimize the number of productive and replacement wells necessary to supply the required steam to the power plants [9, 10]. This method has also been used in other geothermal fields such as Larderello (Italy) [11], Cerro Prieto (Mexico) [12], Wairakei (New Zealand) [13], and Los Azufres (Mexico) [3]. (b) Subsequently, [14] have proposed the harmonic decline curves method, which seems to be more appropriated for geothermal wells data than the decline curves method and hence, it has been widely used. Either by using a mathematical model or a simple qualitative approach, the analysis of the (liquid and steam) well production behavior over time will provide valuable information on the processes occurring at the reservoir.

### 2.2. Analysis of thermodynamic conditions of fluids over time

As soon as the commercial exploitation of a geothermal field starts, some changes in the original reservoir pressure, enthalpy, temperature, etc. distributions take place which are due to both: fluids extraction operations and the occurrence of natural or induced recharge to the reservoir. The magnitude of these variations mainly depends on formation permeability and on the amounts of both the fluids extracted and the recharge. Sometimes the changes in reservoir thermodynamic parameters, as compared with reference or starting conditions, could indicate negative effects in the reservoir, which, in turn, could have an impact on the power generation. Typically, a change in enthalpy indicates a change in reservoir fluid, normally, mass flow also changes with changing enthalpy. This is due to change in pressure drop in the well [7]. If a liquid-dominated reservoir at or near boiling point is exploited, falling pressures induce boiling and the development of two phases at the reservoir. Thus, the higher steam fractions in the fluids entering the well result in higher discharge enthalpies; hence, enthalpies evolution reflects the occurrence and rate of boiling at the reservoir. This is why the monitoring of thermodynamic conditions of fluids entering the wells over time, should be routinely performed.

Usually in two-phase wells, the reservoir temperature can be estimated through chemical geothermometers, being the more important these based on cationic ratios such as the Na/K, [15] and Na-K-Ca [16] and these based on silica solubility [17]. There are some general guidelines in the use of geothermometers that should be considered when choosing the more appropriate geothermometer to estimate the temperature of the stable reservoir in geothermal fields. For example, according to [1], the loss of calcium from the water in calcite precipitation phenomena when the water boils leads to an increase in the Na-K-Ca temperatures. Then, in geothermal fields where calcite deposits occur, the use of the Na/K geothermometer is advisable, since it is independent of boiling mechanisms because it is based on a concentrations ratio rather than on a single concentration. Also, considering that Mg reequilibrates very quickly after changing thermal conditions, the K/Mg geothermometer [18] would not provide the more reliable temperature estimations for the stable reservoir but the temperature at which the last equilibrium occurred.

In steam wells temperatures can be estimated through gas geothermometers [18, 19-23]. According to [18], because of relatively low and similar solubility of both species, the $H_2$-Ar geothermometer can provide reliable reservoir temperatures, while in conjunction with the

$CO_2$-$H_2$ geothermometer, it allows the equilibrium state of the system to be assessed. The gas geothermometers provide temperatures for the fluids at the entrance of the wells, which most often compare well with both, those obtained through wells simulation and those estimated by a silica geothermometer. In order to select the more suitable gas geothermometers to be used for a specific geothermal reservoir, the calculation of those for which data are available is advisable and, then, decision can be made based on contrasting the temperatures obtained with results from other sources (i.e., measured temperatures or simulation temperatures).

### 2.3. Analysis of the chemical composition of water over time

Geothermal fluids usually contain salts in a large range of concentrations. The salinity and the species dissolved in the fluids largely depend on the dominant water–rock interaction processes that take place at high temperatures. The variations of the chemical composition of fluids over time provide important information on the response of the reservoirs to exploitation. For instance, processes like the mixing with cooler groundwaters can be recognized by decreasing trends of both, reservoir temperatures and chlorides (and other solutes) over time. As another example, the process of boiling with limited recharge will increase the discharged fluid enthalpy but decrease the chlorides in total discharge fluids, over time [1, 24-26]. Thus, it is suggested to study the changes in concentrations of ions such as chlorides, sodium, potassium, calcium, sulphate, boron, and silica in both the separated water and the total discharge over time to identify the processes involved.

### 2.4. Analysis of the chemical composition of steam over time

For a long time, the chemical composition of the water has been used to identify the main reservoir physical processes while the steam composition was not fully used due to the sensitivity of gas reactions to pressure changes and also due to factors affecting steam/water ratios or gas solubilities in water, which made difficult the estimation of concentrations of gases at the reservoir. Currently, the understanding of typical effects that reservoir processes cause on the steam has significantly improved [21, 23, 27-29], thus monitoring of steam constitutes a useful tool to investigate reservoir processes. In addition, there are reservoirs in which the water remains immobile and hence only dry steam is produced. For them, the variations of $CO_2$, $N_2$, $H_2S$, total gases, and the $N_2/Ar$ ratio over time could indicate boiling, entry of peripheral fluids, sulphur scaling, etc. If gas/steam ratio, $H_2$, $CH_4$, $CO_2$, and $H_2S$ are available, gas equilibria methods can be used to estimate the reservoir temperatures, in situ steam fraction of fluids entering the wells, and the volumetric liquid saturation at the reservoir [2, 21, 23, 27, 29].

### 2.5. Analysis of $\delta\ ^{18}O$ and $\delta D$ in fluids

As the fractionation of $\delta\ ^{18}O$ and $\delta D$ between two phases is sensitive to temperature changes, isotopes are used to identify processes like boiling, steam separation and condensation in geothermal reservoirs. Thus, the changes in isotopic $\delta^{18}O$ and $\delta D$ composition of fluids produced in geothermal fields are useful to indicate how reservoir processes like mixing, boiling, steam separation, condensation, among others are progressing during exploitation.

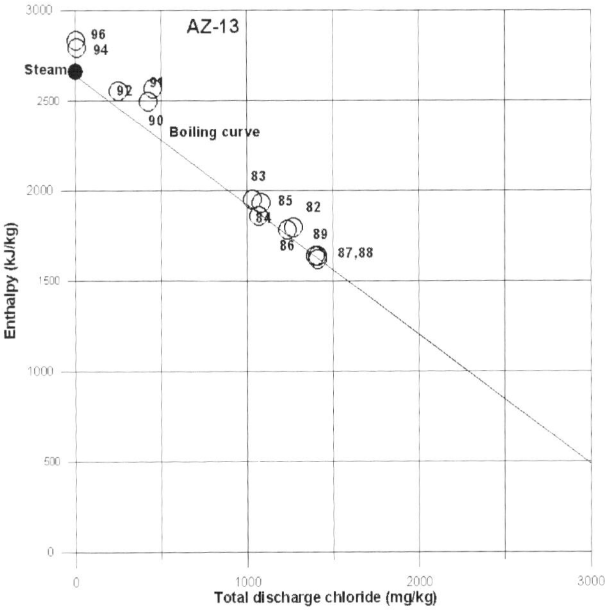


Figure 2. Annual average enthalpy vs total discharge chlorides for well AZ-13.

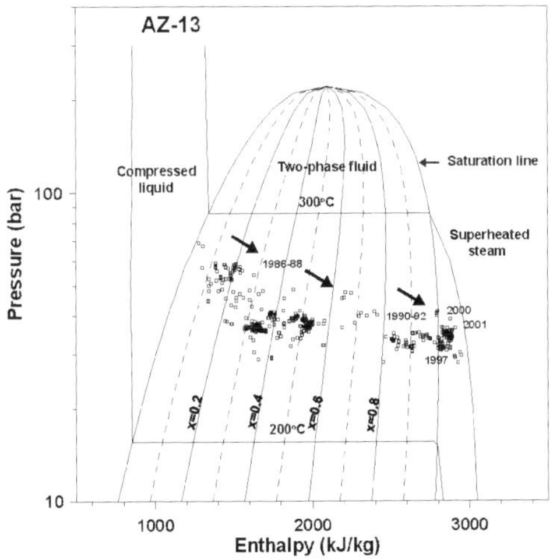


Figure 3. Well-bottom pressure vs well-bottom enthalpy for well AZ-13. The rows indicate evolution of thermodynamic conditions of fluids entering the well, from two-phase to steam.

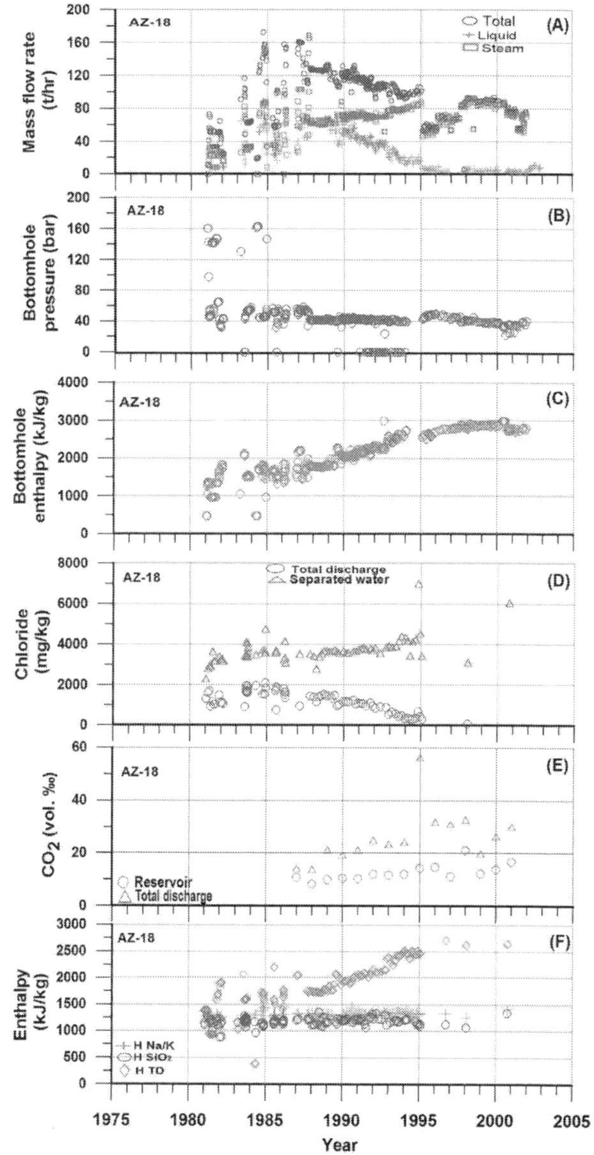


**Figure 4.** Time series of (A) mass flow rates, (B) bottom-hole pressure, (C) bottom-hole enthalpy, (D) chlorides in total discharge and separated water, (E) $CO_2$ in total discharge and in the reservoir, and (F) total discharge, Na/K and silica estimations of enthalpy for well AZ-18.

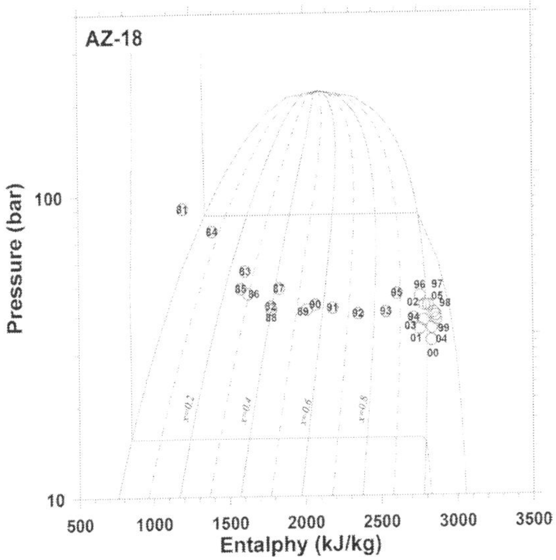


**Figure 5.** Well-bottom pressure vs well-bottom enthalpy for well AZ-18.

## 3.2. Cooling and mixing processes

The pressure difference between hot and cold aquifers increases as a result of exploitation and this situation can induce the entry of cooler waters from neighboring aquifers to geothermal reservoirs. The cooler water entry process can be identified by geochemical changes of the fluids produced since most often cooler waters are more diluted than hot geothermal fluids then dilution effects are observed over time. As an example, the case of the Las Tres Vírgenes geothermal field [6] is given. In order to identify reservoir processes for the wells and to define the time periods in which each process was dominant, the comparison of the total discharge, Na/K and silica enthalpies according to the method proposed by [34] was used. In Figure 6, the total discharge enthalpy and the enthalpies from silica and Na/K geothermometers for well LV-4 are given. In this figure, important variations in the total discharge enthalpy from 2001 to 2004 indicate more than one fluid entry to the well and also near-well boiling of the fluids flowing to the well, in response to decrease in well-bottom pressure. As a result of near-well boiling, the total discharge enthalpy increases. At the same time, the inflow of lower temperature fluids to the well is identified by the pattern $H_{SiO2} > H_{Na/K}$, because the Na/K geothermometer is a "slow response" geothermometer compared with $SiO_2$ geothermometer, thus indicating a previous temperature of the fluid. In 2004, the pattern $H_{TD} \cong H_{Na/K} \cong H_{SiO2}$ indicates that the well was producing from an equilibrated liquid. Because of the relatively low total discharge enthalpy, it is inferred that the near-well boiling process has stopped. During 2005, the pattern observed is $H_{SiO2} > H_{Na/K} = H_{TD}$, which indicates a thermal breakthrough of cooler water with Na/K enthalpy reequilibration. This unusual pattern could have been induced by reduction in the production orifice, since the well was being closed at that time. During 2006, the well started production from an equilibrated liquid. In 2007, the enthalpy pattern $H_{TD}$ = $H_{Na/K}$ > $H_{SiO2}$ shows mixture close to the well of cooler, more diluted water with equilibrated liquid. Lower silica enthalpies result from dilution without silica reequilibration. In well LV-11 (Figure 7) during 2006 the pattern $H_{Na/K}$ > $H_{SiO2}$ > $H_{TD}$ indicates very clearly the mixture in the well of cooler, more diluted water with equilibrated liquid, while during 2007 the pattern $H_{Na/K}$ > $H_{SiO2}$ = $H_{TD}$ results from mixing with cooler water near the well with reequilibration of silica enthalpy but not of Na/K enthalpy.

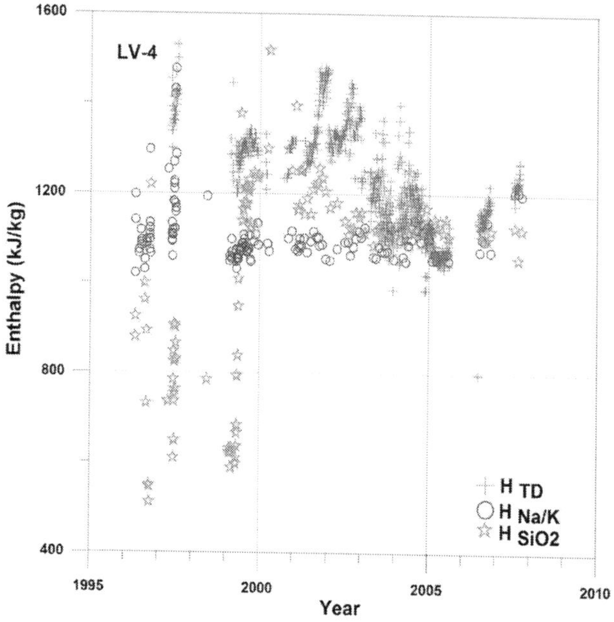


**Figure 6.** Time series of total discharge, Na/K and silica enthalpy estimations for well LV-4.

The enthalpy histories for well LV-13 are given in Figure 8. The enthalpy comparison pattern observed – $H_{TD}$ > $H_{SiO2}$ >$H_{Na/K}$ – indicates near-well boiling for short periods of time, which corresponds to increases in total discharge enthalpy, and because of the constant pattern observed (up to 2006) – $H_{SiO2}$ >$H_{Na/K}$ – mixing with cooler water near the well also occurs.

Based on the investigation of reservoir processes, apart from short time near-well boiling processes found for wells LV-4 and LV-13, it could be said that the more important processes that were identified to occur in the studied wells were the entry of lower temperature waters and the multiple entries of fluids to the wells for relatively long periods of time. However, such processes were not severe in well LV-11, since during 2003-2004 it showed a stable behavior.

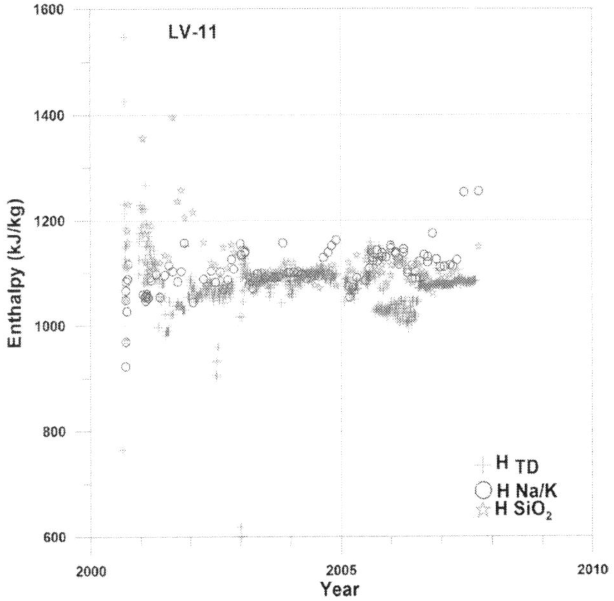


Figure 7. Time series of total discharge, Na/K and silica enthalpy estimations for well LV-11.

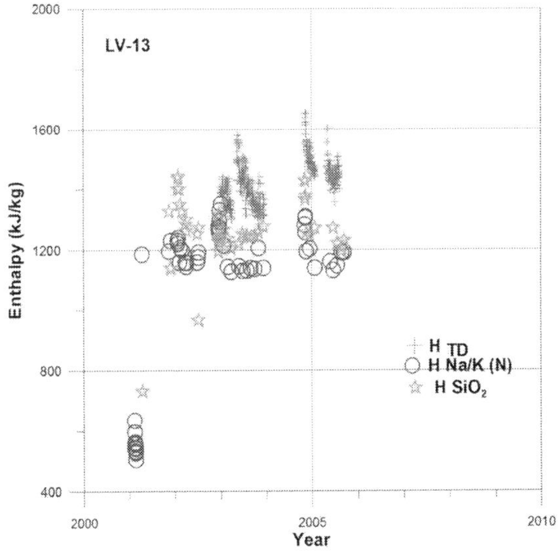


Figure 8. Time series of total discharge, Na/K and silica enthalpy estimations for well LV-13.

From the enthalpies comparison method used in the Las Tres Vírgenes wells, it is clear that more diluted and cooler water with respect to the reservoir fluid is entering the reservoir because of exploitation. In order to estimate the enthalpy of the diluting end-member involved in the mixing process, the enthalpy-chloride model [37] was used [6]. To follow this approach, annual average Na/K enthalpies versus annual average total discharge chlorides for the wells were plotted in Figure 9. As seen in the figure, a linear tendency of the data indicates that the reservoir fluids which originally plot on the boiling line with an average reservoir (Na/K) enthalpy of 1203 kJ/kg (274°C) and chloride concentration of 6,745 mg/kg become cooler and more diluted over time because of mixing. For the diluting water, an enthalpy of 798 kJ/kg (187°C) and a chloride of ~500 mg/kg were estimated.

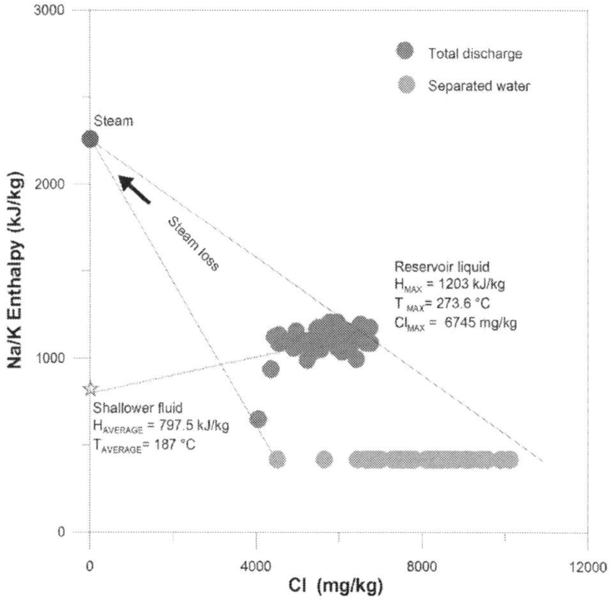


**Figure 9.** Enthalpy vs chloride mixing model.

The estimated chloride of the diluting water can be found by lineal relationships between two conservative components as B versus Cl, given in Figure 10, considering well compositions. In such relationships, the diluting end-member was the fluid for the well LV-2, which was the shallower well drilled in the field and for which temperature estimations agreed with the temperature found for the diluting end-member from the enthalpy-chloride model. At present, well LV-2 is no longer productive; therefore, new samples are not available to confirm these results. From this study it was proposed that the severe scaling phenomena observed in the field could be attributed to both the chemical nature of the fluids and the mixing of different fluids, since mixing phenomena strongly enhances scaling [24-26, 38].

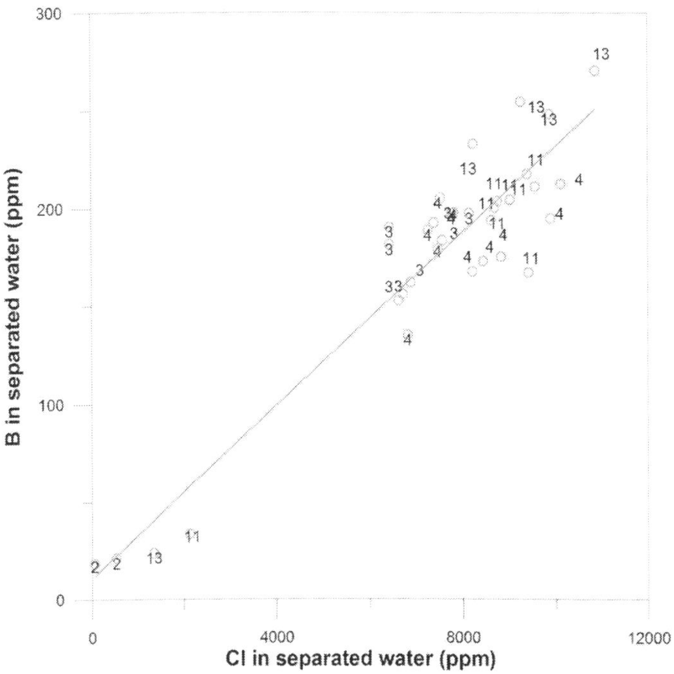


**Figure 10.** B vs Cl in separated water for the Las Tres Vírgenes wells.

### 3.3. Production of reinjection returns

The historical data on mass flow rates of fluids reinjected to the reservoir through specific reinjection wells could be related to production and geochemical data of production wells to assess the influence of reinjection. Isotope and chemical data of reinjection fluids can also be related to characteristic geochemical changes in fluids produced by wells affected by reinjection.

In order to investigate if reinjection fluids have arrived to the production zone of well AZ-13 of the Los Azufres geothermal field, production data of the well AZ-13 were analyzed together with histories of reinjection wells and isotope data for both, the producer well and the injection fluids [5]. As seen in Figures 1B and D in 1995, the mass flow rate produced by the well AZ-13 slightly increased while enthalpy slightly decreased. Both effects occurred when the mass flow rate of water injected into well AZ-15 (Figure 11) was at its maximum level. The distance between these wells is about 3 km. When the injection mass flow rates decreased, the mass flow rate of well AZ-13 also decreased and the enthalpy increased. Variations in δD of fluid discharged from well AZ-13 also follow the pattern of injection into well AZ-15 (Figure 11); the fluid produced by well AZ-13 was depleted in δD when injection rates were low and was enriched when the rate was high. This behavior occurs because the reinjected fluid becomes isotopically enriched by evaporation at the surface. These observations suggested that some of the injected fluid in well AZ-15 arrived at the production zone of AZ-13, although the amount was not high enough to produce significant changes in the thermodynamic conditions of the fluid that feeds well AZ-13. In general, the behavior described for the well AZ-13 is typical of the wells in the north zone of the Los Azufres geothermal field. The reinjection effects in Los Azufres geothermal field are more important in the south zone wells [5].

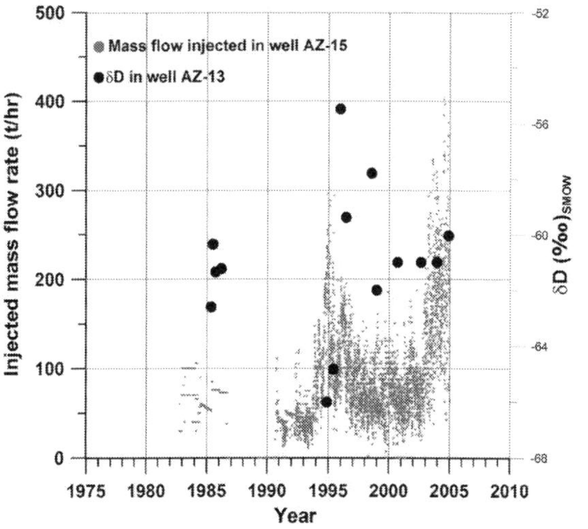


**Figure 11.** Mass flow rate injected in well AZ-15 and δD in the total discharge of well AZ-13 over time.

Clear evidence of the production of reinjection returns was obtained for well AZ-2 located in the south zone of the Los Azufres geothermal field; this well behavior was studied by [4]. Well production and estimated well-bottom parameters for well AZ-2 over time are given in Figure 12.

Production data for this well consisted of two stages: from April 1978 to April 1984 and from May 1988 to April 2001. During the first stage of production, an irregular behavior of production (Figure 12A) and in estimated well-bottom variables, as pressure (Figure 12B) and enthalpy (Figure 12C) is observed. By then, injection in wells AZ-7 and AZ-8 just started; thus, reinjection was not the only event affecting well behavior. Such irregular behavior seems to be related to mechanical damage in the pipe. For the second production stage, the well still showed a very irregular behavior, but in this case, an enrichment in chlorides since May 1988 (Figure 12D) and a decrease in Na/K enthalpy for the feeding fluid regarding these for silica and for the total discharge is indicated (Figure 12E). Also, the variations in well-bottom pressure (Figure 12B) and well-bottom enthalpy (Figure 12C) are in opposite directions. For pressure increments enthalpy decreases are seen. Pressure increments are related to recharge

http://dx.doi.org/10.5772/61586

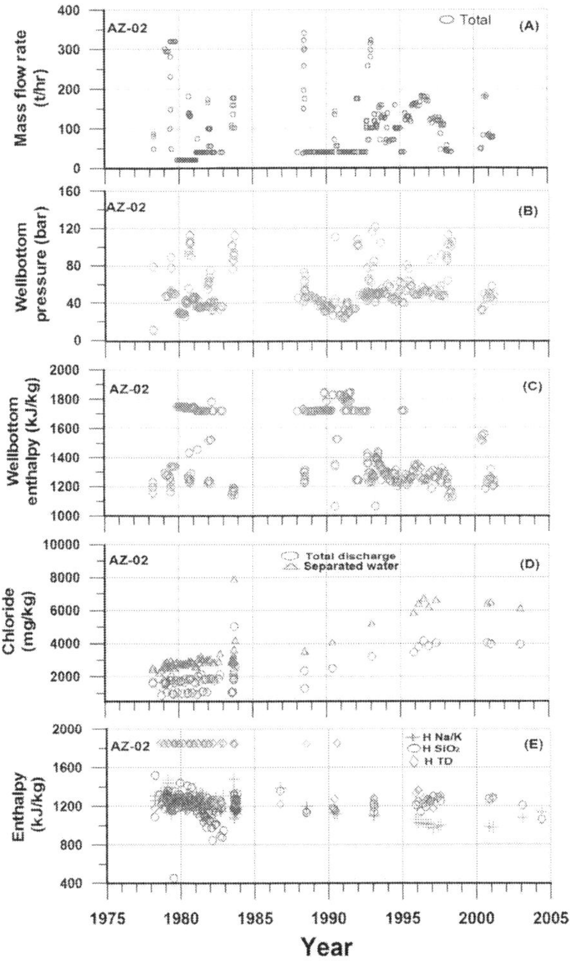


**Figure 12.** Time series of (A) mass flow rates; (B) well-bottom pressure; (C) well-bottom enthalpy; (D) chlorides in total discharge and separated water; and (E) total discharge, Na/K and silica enthalpy estimations for well AZ-2.

processes while decreases in enthalpy indicate that the recharge is a lower enthalpy fluid as compared with that of reservoir fluids. The well production also shows large variations (Figure 12A), which are related to the amount of recharge arriving to the production zone of the well. At the same time, the mass flow rate (Figure 12A) and the $\delta^{18}$ O of well AZ-2 follow the same pattern as the mass flow rate injected in well AZ-7 (Figure 13). These reinjection effects could not be associated with well AZ-8 because of significant reinjection in this well that started in 1990. Thus, it could be concluded that during the second production stage, the feeding fluid of well AZ-2 was being gradually replaced by another fluid having higher salinity and lower temperature than original fluid. It seems feasible that this fluid comes mainly from injection in well AZ-7. This cooler fluid during flowing to the well AZ-2 is heated in contact with the reservoir rock; however, it is not heated enough and the well produced a relatively cooler fluid.

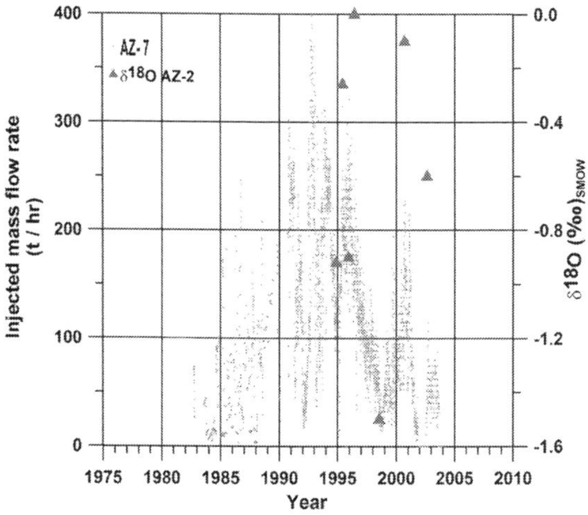


**Figure 13.** Mass flow rate injected in well AZ-7 and $\delta^{18}$ O of the total discharge of well AZ-2 over time.

## 4. Conclusions

Exploitation of geothermal resources produces changes in geochemical and production data of wells regarding natural-state conditions, due to the extraction (and sometimes the reinjection) of fluids. These changes can be used to investigate the main physical processes in the reservoir as a result of exploitation through the analysis of monitoring data of wells. A methodology based on the variations of geochemical and production data of wells over time was described. This innovative, combined methodology provides more reliable estimations of reservoir processes than current techniques based either on geochemical or production data, since the thermodynamic conditions of well-bottom fluids are included into analysis. In this regard, the main tendencies of well-bottom pressure and enthalpy data obtained from simulation of wells in suitable diagrams indicate how exploitation impacts the reservoir and help delineating future strategies. This approach can be used not only for single wells but also to study groups of wells to investigate zones of the reservoir affected by a specific process. The methodology has been successfully used to investigate the response of Mexican geothermal reservoirs to exploitation, in which the main processes identified included mixing, boiling, and production of reinjection returns in production wells, and examples for them have been provided. The results from this approach together with the results from other disciplines provide support in field management on delineating optimal exploitation strategies to prolong the reservoir lifetime.

Chapter 5

# Airborne Magnetic Surveys to Investigate High Temperature Geothermal Reservoirs

Supri Soengkono

Additional information is available at the end of the chapter

http://dx.doi.org/10.5772/61651

**Abstract**

Airborne magnetic survey is an effective geophysical exploration method in terms of coverage, resolution and cost, particularly for area with restricted or difficult ground access. Research studies in New Zealand have shown airborne magnetic surveys can indicate the regions of high reservoir permeability and thermal up-flow of active geothermal systems. However, the method has not been extensively used in the geothermal investigations, probably because the interpretation of airborne magnetic data has so far been seen as difficult and requires a complex quantitative 3D modelling of subsurface magnetisation.

This chapter introduces a new approach to use airborne magnetic survey to investigate high temperature geothermal resources without the need of 3D magnetic modelling. This new approach takes advantage of data processing packages that during the last few years have become accessible through the internet. A simple but comprehensive explanation is given on the physics background of the airborne magnetic surveys. Examples are provided from interpretations of real airborne magnetic data from the North Island of New Zealand and the Java Island of Indonesia. This chapter is aimed to provide the readers a sufficient level of knowledge and confidence to organise and/or run investigation of high temperature geothermal reservoirs using airborne magnetic surveys.

**Keywords:** Total force magnetic anomalies, magnetisation of volcanic rocks, hydrothermal alteration and demagnetisation, extent of geothermal reservoirs, reversely magnetised rocks

## 1. Introduction

Airborne magnetic survey involves measurements of the geomagnetic field (the magnetic field of the earth) from the air using magnetometer installed in an aircraft. The purpose is to detect small changes in the geomagnetic field related to differences in rock magnetisation beneath the survey area. The airborne magnetic survey has been extensively applied in the mineral explorations and as additional tool to support geological mapping projects. It is a very effective geophysical exploration technique in terms of coverage, resolution and cost, particularly in area with difficult or restricted ground access. The airborne magnetic surveys over the Taupo Volcanic Zone (TVZ) in New Zealand, which started with a regional survey dating back more than 60 years ago in the early 1950s [9] and was followed in much later date by a variety of more detailed surveys between 1984 and 2006 [24,14,31], have provided data sets that all are highly consistent to each other. The resolution of the dataset, as expected, depends on the flight line spacing and the survey altitude above the ground. But all the data sets revealed the same features of magnetic anomalies. This consistency clearly shows that airborne magnetic survey is a robust geophysical method. Over geothermal prospect regions worldwide, airborne magnetic data are often already available from some previously conducted surveys by mineral exploration companies searching for epithermal gold deposits or by the government institutions (Geological Surveys).

Hochstein and Soengkono [14] showed that careful interpretations and three dimensional (3D) quantitative modelling of airborne magnetic anomalies over many geothermal systems in the Taupo Volcanic Zone (TVZ) in New Zealand can provide information on the likely locations of high reservoir permeability and up-flow regions of active geothermal system. They also quoted examples that suggest this would also likely be true for geothermal systems in volcanic settings elsewhere outside New Zealand. However, the interest to use data from airborne magnetic surveys to investigate high temperature geothermal systems has been very slow to develop. A possible reason of this slow development is that the interpretation of magnetic data in geothermal investigations has been considered difficult and the 3D quantitative modelling of the magnetic data can be a complex and problematic task. The changing in geomagnetic inclination at different geographic latitudes causes different pattern of magnetic anomalies over areas which have the same geological structures and lithology but are located at different geographic regions. In addition, unlike the scalar parameter rock density that causes gravity anomalies, the rock magnetisation is a vector. Because of this, the pattern of the magnetic anomalies is complex and more difficult to interpret than the gravity anomalies over the same area. However, with the development of user friendly geophysical processing packages that, since mid-2000s, have become accessible on-line (worldwide), the complexity of magnetic anomalies can now be reduced.

This chapter introduces and explains a new approach to use airborne magnetic data for the investigations of high temperature geothermal resources hosted by volcanic rocks. This new approach is specifically formulated for this chapter based upon the author's experience during the last 30 years in the interpretation and 3D modelling of various airborne magnetic data. It has not been previously published in any papers listed in the reference list of this chapter (Section 7), nor anywhere else. This approach utilises some magnetic data processing techniques in the computer software that have now become accessible worldwide. The processing techniques are used to directly link the measured airborne magnetic anomalies to the causative source targets. The often complex and difficult 3D modelling of the anomalies would only need to be carried out when it is considered necessary at the final stage of the interpretation, when some further detailed aspects of the magnetic interpretation need to be pursued. The aim of this chapter is to equip readers with some knowledge and confidence to run the investigation of high temperature geothermal reservoirs in volcanic rocks using airborne magnetic data which are already available over target area from some previous surveys, or going to be collected by a new survey specifically aimed to explore the geothermal targets.

The SI unit of magnetic field strength is Tesla, or T (T=Weber/$m^2$=Vs/$m^2$). The unit used for geomagnetic field strength measured during airborne magnetic survey is nT (1 nT = $10^{-9}$T), which is equal to the old unit gamma ($\gamma$) (1nT = 1$\gamma$).

## 2. The geomagnetic field

### 2.1. The normal geomagnetic field

The normal (undisturbed) geomagnetic field $\mathbf{B}_0$ can be approached by the effects of a fictitious magnetic dipole at the centre of the earth, orientating at a small angle (about 10°) to the axis of earth rotation. On the surface of the earth, the *inclination* of the normal geomagnetic field causes by such a dipole varies from +/-90° at the magnetic north and south poles, to 0° at the magnetic equator (see Figure 1). As the positions of the north and south magnetic (N′ and S′) and geographic (N and S) poles are not the same, there is also a horizontal *declination* between the magnetic north (the horizontal direction of the earth's magnetic field shown by a compass needle) and the geographic north. The magnetic inclination is considered positive downwards; it is negative in the southern hemisphere (such as in NZ, Australia and Africa). Since the fictitious dipole at the centre of the earth is only quasi static (it has a slow precession around the earth rotational axis), there is a small secular variation of the normal geomagnetic field.

Studies of *remanent* magnetisation (the permanent magnetisation that is not induced by the present day geomagnetic field) of the rocks of different ages from around the earth show that the orientation of the fictitious dipole at the centre of the earth had flipped many times in the past, when the positions of N′ and S′ were interchanged (called the geomagnetic reversal). The last geomagnetic reversal occurred about 0.7 Myr ago [19].

The normal geomagnetic field of the earth ranges in strength from about 35,000 nT near the equator to about 60,000 nT near the north and south poles. For examples, in the North Island of New Zealand (about 38°S latitude) it has the strength of about 54,000 nT whereas in the Java Island of Indonesia (about 7°S latitude) the geomagnetic field strength is about 44,000 nT. At any point on the surface of the earth, the strength and direction (declination and inclination) of the normal geomagnetic field can be obtained from the internet (by searching for "geomagnetic field strength" to get to web sites to do the calculation online). The geomagnetic values are computed using a global model, defined by spherical harmonic coefficients synthesising the quasi static component as well as the secular variation of the earth's magnetic field, and is known as the International Geomagnetic Reference Field (IGRF). This global model is renewed every about 5 years by collaborative effort between magnetic field modellers and the institutes involved in collecting and disseminating magnetic field data from satellites and from observ-

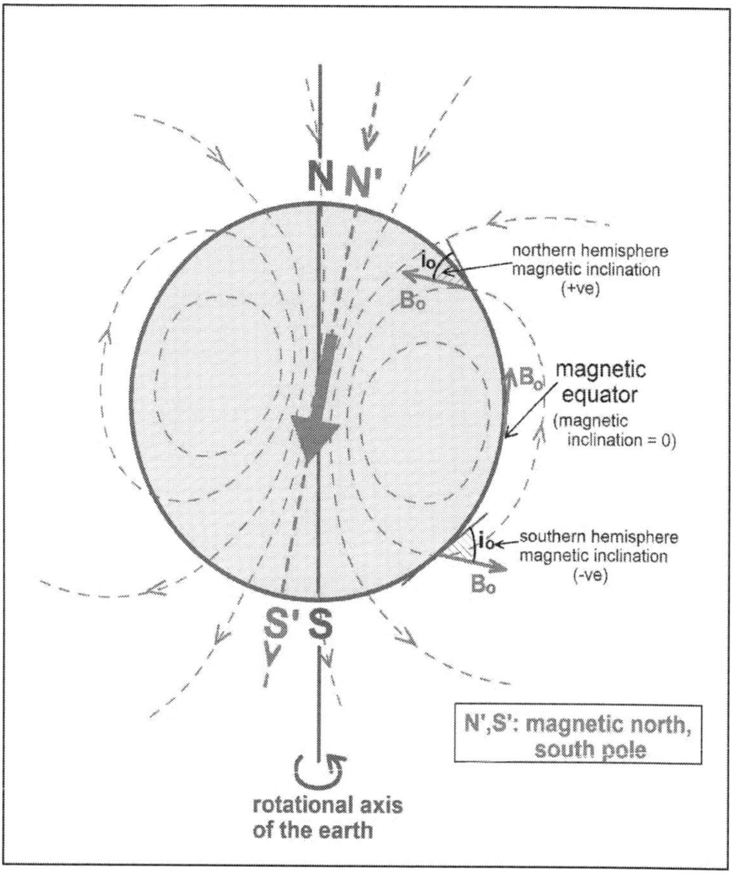


**Figure 1.** The normal geomagnetic field of the earth.

atories and surveys around the world. Note that the value of the magnetic inclination is not the same as the value of geographical latitude.

### 2.2. Diurnal variations

The total geomagnetic field measured during a magnetic survey contains small magnetic fields related to local variation of rock magnetisation (this is the target of airborne magnetic survey) and time variant external components from outside the solid earth. The time variant external field includes small diurnal variations (range about 30 nT) of about 24 hours period which correlate with electrical currents in the ionosphere, and a larger transient and erratic disturbance (range up to 1000 nT) known as magnetic storms which correlate with sunspot activity. The effect of the time variant external fields has to be corrected from the measured airborne

http://dx.doi.org/10.5772/61651

magnetic data. If magnetic storm is avoided when conducting the survey, the only significant time variant field affecting the measured geomagnetic field would be that causing the diurnal variations. Near the earth's surface, the magnetic field causing the diurnal variations has an almost constant strength and direction within a 75 km radius. For a magnetic survey it is therefore sufficient to monitor the diurnal variations at a base station within or near the survey area. Data provided by airborne magnetic survey companies contracted to carry out the survey would have been corrected for this diurnal variation.

### 2.3. Magnetic anomalies

Figure 2 shows a schematic vector diagram of the relationship between total geomagnetic field ($\mathbf{B}_{obs}$), the normal (undisturbed) geomagnetic field ($\mathbf{B}_o$) and the total sum of local magnetic fields produced by variations of rock magnetisation ($\mathbf{\Delta B}$), at two theoretical measurement points (a) and (b) having two different $\mathbf{\Delta B}$ vectors. The strength of total geomagnetic field vector ($=|\mathbf{B}_{obs}|$) is the *only* parameter measured during airborne magnetic survey. As shown in Figure 2, this measured geomagnetic magnetic field $\mathbf{B}_{obs}$ vector is the result of vector operation $\mathbf{B}_{obs} = \mathbf{B}_o + \mathbf{\Delta B}$. The sum magnetic field strength caused by anomalous bodies ($|\Delta B|$) is usually much smaller (less than 2%) compared to the strength of the normal geomagnetic field ($|\mathbf{B}_o|$) (except at a ground location close to an outcrop of very highly magnetic rocks). Hence, the angle between $\mathbf{B}_o$ and $\mathbf{B}_{obs}$ is very small (less than $2^{\circ}$).

The *total force magnetic anomaly* (termed in this chapter as $\Delta F$) is defined by algebraic (scalar) subtraction of $|\mathbf{B}_o|$ (the strength "undisturbed" earth's magnetic field) from $|\mathbf{B}_{obs}|$, that is: $\Delta F = |\mathbf{B}_{obs}|-|\mathbf{B}_o|$. As $|\mathbf{B}_o|$ cannot (are not) directly measured, the value $|\mathbf{B}_o|$ it is usually taken from computation of the IGRF (see Section 2). The $\Delta F$ computed from the measured $|\mathbf{B}_{obs}|$ subtracted by the IGRF value contains all the magnetically anomalous mass beneath the measurement point down to very deep level (theoretically it is down to the depth of the fictitious magnetic dipole near the centre of the earth). To approximate $\Delta F$ affected only by magnetically anomalous mass beneath the "survey target", one can use the trend of $|\mathbf{B}_{obs}|$ across the survey area to determine $|\mathbf{B}_o|$. No exact value of such survey target "depth" can be given, but the $\Delta F$ values obtained in this way are better for identifying and delineating the survey target. For this reason, computer software that can properly grid airborne magnetic data and compute the trend of gridded data is essential in the interpretation of airborne magnetic surveys. The computation of trends and other processing and plotting of the magnetic data presented in this chapter are all carried out using the computer software Oasis Montaj from the Geosoft Inc. There are some other software packages available in the market (can be purchased online) that can do the same processing and plotting.

If the airborne magnetic data that available for the investigation are already in the form of $\Delta F$ obtained by subtracting of the IGRF values (often this is the case with data received from the airborne magnetic survey contractor), removing the trend of such $\Delta F$ values over the survey area will provide new (corrected) $\Delta F$ values that would represent magnetically anomalous bodies no deeper than the "survey target".

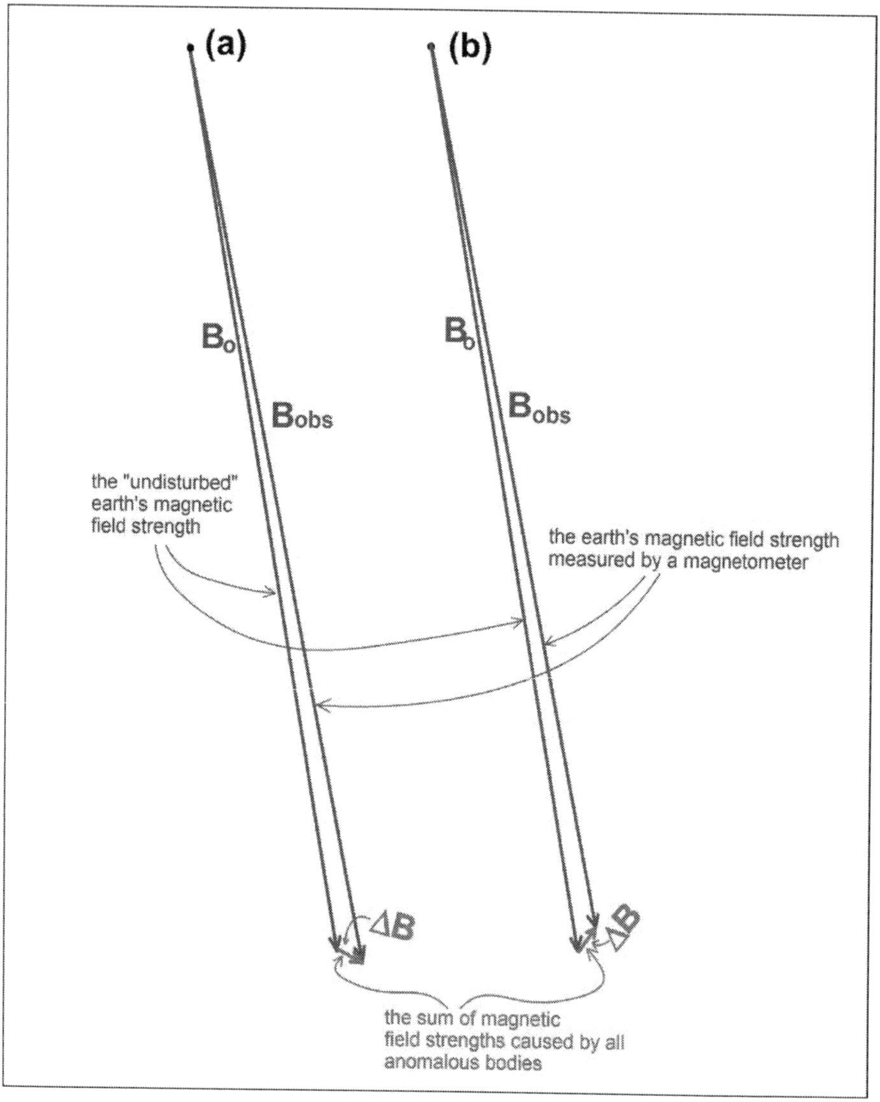


**Figure 2.** Vector diagram of magnetic anomalies. In (a) $\mathbf{\Delta B}$ adds to $\mathbf{B}_o$ and we obtain positive total force magnetic anomaly $\Delta F$. In (b) $\mathbf{\Delta B}$ opposes $\mathbf{B}_o$ and the total force magnetic anomaly is negative.

http://dx.doi.org/10.5772/61651

### 2.4. The patterns of a magnetic anomalies caused by magnetic dipole

First order effects of observed magnetic anomalies can often be interpreted by simple dipole fields (such field are set up by geological bodies that can be approximated by a magnetic dipole or combination of dipoles). This is especially true for geological bodies that can be approximated by a homogeneous sphere, since homogeneous sphere is magnetically equivalent to a single magnetic dipole placed at its centre.

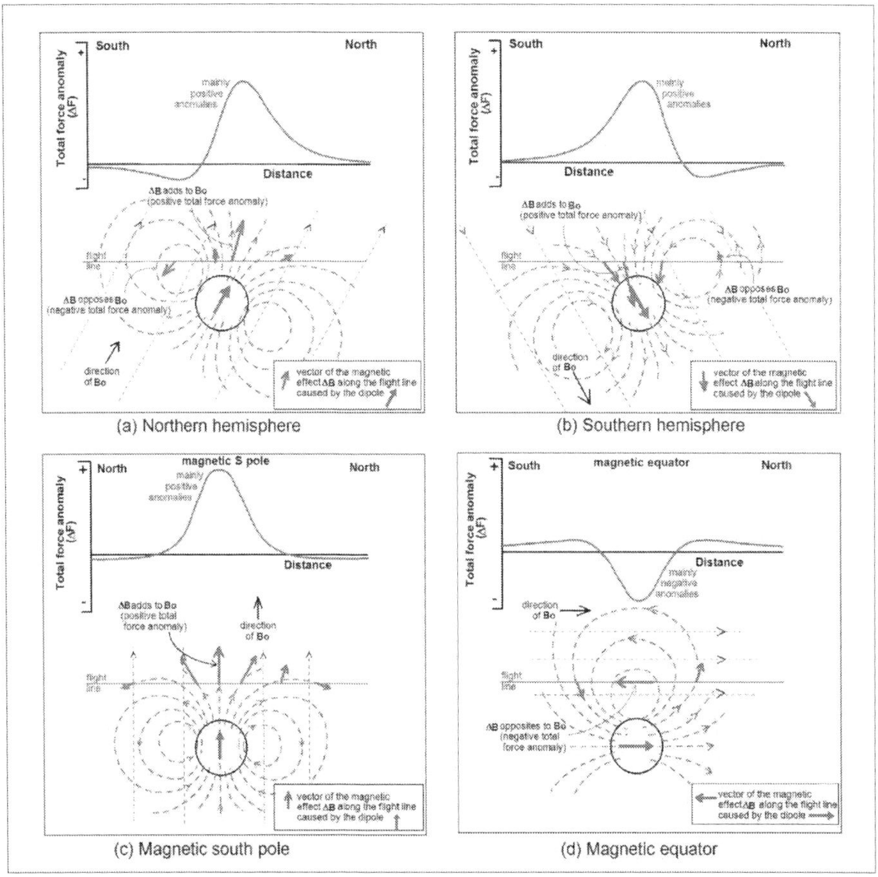


**Figure 3.** Total force magnetic anomalies caused by a magnetic dipole (or a magnetic sphere) at different geomagnetic inclination (different geographic latitudes).

Figure 3 shows diagrams explaining the patterns of total force anomalies ($\Delta F$) caused by *positive* magnetic dipoles (orientated in *the same direction* as $\mathbf{B}_0$) that are located in: (a) the northern hemisphere, (b) the southern hemisphere, (c) the magnetic *south* pole and (d) the magnetic equator. For the magnetic *north* pole, by using a diagram equivalent to Figure 3 (c) but with a vertical downward direction of the magnetic dipole and $\mathbf{B}_0$, it can easily be shown that the pattern of the created total force anomaly (ΔF) is exactly the same as that shown in Figure 3 (c). The same can be made also for *negative* magnetic dipoles (orientated *opposite in direction* to $\mathbf{B}_0$), which will show the patterns of total force anomalies in Figure 3, but with ΔF values having the opposite sign. Figure 3 also shows that total force magnetic anomalies ΔF created by the *positive* magnetic dipoles at locations away from the magnetic equator are dominantly *positive*. At the south (or north) magnetic pole (Figure 3(c)), the centre of positive anomalies is located directly above the dipole. At the southern hemisphere (Figure 3(a)) the centre of the positive anomaly is shifted to the north of the magnetic dipole and at the northern hemisphere (Figure 3(b)) it is shifted to the south of the magnetic dipole. However, the opposite occurs at the locations close (within about ±10°) to the magnetic equator. Here, the magnetic anomalies created by a *positive* magnetic dipole become dominantly *negative*! This "counter intuitive" phenomenon can result in a serious error in the interpretations of magnetic anomalies at locations within about ±10° magnetic latitude as at some parts of Africa, South America and the southern part of the Philippines.

## 3. Interpretation of magnetic anomalies over geothermal areas

### 3.1. Magnetic anomalies over the Taupo Volcanic Zone (TVZ) in New Zealand

The Taupo Volcanic Zone (TVZ) is a region of Quaternary volcanic and geothermal activity in the central North Island of New Zealand. Almost all high temperature geothermal systems in New Zealand are located in this zone. As mentioned in the introduction, a regional airborne magnetic survey was conducted over the TVZ during early 1950s and it was followed much later on by more detailed surveys between 1984 and 2006. The geomagnetic field inclination in the TVZ is about -62.5°. Because of this high inclination, the relationship between magnetic anomalies and geological features is easier to recognise than in the regions with lower magnetic inclination. The phenomenon that distinct magnetic anomalies are often associated with geothermal reservoirs in the TVZ has been recognised more than 70 years ago by Watson-Munro [35]. However, in the regional map drawn using the data from the early 1950s regional survey (for example, Hunt and Whiteford [16]) the relationship between some geothermal fields and the magnetic anomalies is not obvious and can just be barely seen because of wide flight line spacing (2.5 km) of the survey. The interest in using magnetic surveys as an exploration tool declined in New Zealand in the 1960s when electrical methods were found to be more effective in delineating the lateral extent of conductive reservoir rocks.

The interest was revived when the Geothermal Institute (University of Auckland, New Zealand) started new, lower-level airborne magnetic surveys in 1984. One of the first surveys led to the discovery of a distinct magnetic anomaly over the Mokai geothermal field [20] that was hardly recognisable in the low resolution early map from the early 1950s survey. Between 1984 and late 1999s, the Geothermal Institute expanded the survey to cover an area of about 3000 km$^2$, covering all geothermal fields in the TVZ. Quantitative interpretation of more than

10 geothermal prospects within the area were undertaken [20,14,12-13,26-27,22-23,28-29,21,25,15]. The quantitative 3D interpretations indicated that airborne magnetic survey can be a useful tool to identify zone of high permeability region of an active geothermal system.

### 3.2. Magnetisation volcanic rocks

All volcanic rocks were magnetic after their eruption, as a result of their induced magnetisation ($\mathbf{m}_i$) and remanent magnetisation ($\mathbf{m}_r$). Remanent magnetisation is the permanent magnetisation of rock which was attained when the rock was formed (or deformed). The main type of remanent magnetisation in volcanic rocks such as lavas and ignimbrites is the thermo remanent magnetisation (TRM) which was attained when the rocks cooled down to below the Curie point of magnetite (about 580 °C). It has the same direction as the past $\mathbf{B}_o$ during the time of cooling. Induced magnetisation is given by the multiplication product of the geomagnetic field magnetising force $\mathbf{H}$ (A/m) and the magnetic susceptibility $\kappa$ (a dimensionless parameter) which, in turn, is related mainly to the volume fraction of two primary magnetic minerals, magnetite and titanomagnetite. In the SI unit, the earth's magnetising force $\mathbf{H}$ can be determined from the relationship $\mathbf{B}_o = \mu$ H, where $\mu$ is the magnetic permeability of the medium. For non-magnetic medium (such as air) $\mu = 4\ \pi\ x10^{-7}$ In the magnetic survey, the unit of $\mathbf{B}_o$ is nT (=$10^{-9}$T). Thus, H (in A/m) can be obtained from $\mathbf{B}_o$ (in nT) from $H=(1/\mu)\mathbf{B}_o$.= $(10^{-9}/(4\pi x10^{-7}))\mathbf{B}_o = 7.96x10^{-4}\mathbf{B}_o \approx 8x10^{-4}\mathbf{B}_o$ Hence, the induced magnetisation is given by the equation: $\mathbf{m}_i = \kappa\mathbf{H} = \kappa(8x10^{-4})\mathbf{B}_o$. $\mathbf{m}_i$ has the same direction as the present day $B_o$.

In the TVZ, the strength of the induced magnetisation ($|\mathbf{m}_i|$) of rhyolites and ignimbrites is of the order of 0.5 A/m [21], pointing to the presence of about 0.8% (by volume) of primary magnetic minerals [17]. Petrology studies by Ewart [8] indicate magnetite values between 0.3 and 0.8% from "point counting". The strength of the remanent magnetisation ($\mathbf{m}_r|$) of these rocks is significantly greater and lies commonly within the range of 1-4 A/m [21]. The strength of the total magnetisation ($\mathbf{m}_t = \mathbf{m}_r + \mathbf{m}_i$) of unaltered, normally magnetised volcanic rocks in the TVZ lies between about 0.5 A/m (tuffs, pumice, and volcanic breccia) and about 2.5 A/m (rhyolite lavas) [14]. The values cited are median values obtained from more than 200 rock samples measured by students at the Geothermal Institute (University of Auckland) between 1988 and 1993. The effect of an overprinted events associated with the natural fluctuation of geomagnetic field (known as the viscous remanent magnetisation, or VRM) in young rocks such as Quaternary volcanic rocks is likely to be small and can be neglected.

### 3.3. Hydrothermal demagnetisation of volcanic rocks

Hydrothermal alteration usually causes a "de-magnetisation" of initially magnetic reservoir rocks. Hydrothermal demagnetisation causes *negative* magnetisation contrast. Petrology studies show that, in these reservoirs, primary (titano-) magnetite has been replaced by almost non-magnetic minerals such as pyrite, leucoxene, or hematite [32]. In New Zealand some petrology studies in the Geothermal Institute (University of Auckland) showed that in liquid-dominated systems (titano-) magnetite appears to be the first mineral replaced during thermal alteration; this also applies to many liquid dominated systems in the Philippines and Indonesia

(Prof P.R.L. Browne, pers. comm., 1994). Hydrothermal demagnetisation appears to also occur in exposed rhyolite domed near the margins of some TVZ geothermal systems [20,26]. It seems acidic condensate formed in shallow vapour zones can cause demagnetisation of volcanic rocks forming topographic highs.

The solubility of magnetite increases rapidly in aqueous solutions with decreasing pH values (pH < 6) and at temperatures below 200°C [4]. Magnetite is also unstable in volcanic rocks that are saturated with $CO_2$-rich, steam heated waters with a pH value of <6. These occur, for example, near the top and margins of the Ohaaki reservoir in the TVZ [11]. The waters are corrosive and have caused significant external corrosion to well casings. All cores from the cooler well at Ohaaki which discharged such fluid were nearly non-magnetic. Since cooler, $CO_2$-rich waters at the margin of a gas-rich field like Ohaaki can also occur outside its boundary, it is possible that non-magnetic rocks can extend beyond the boundary of such a field delineated by resistivity surveys. Magnetite can be stable in the deeper oxidising environment of the natural two-phase system of Olkaria (Kenya) and, in that reservoir, it is less affected by thermal alteration than all other primary phases except quartz (Prof P.R.L Browne, pers. comm. 1994); the same applies to host rocks of the El Tatio outflow in Northern Chile. Because of this, the airborne magnetic survey identifies only the upper part of the geothermal reservoir. Except in the situations where the ground water flows have very strong horizontal component (associated, for example, with steep overall topography) the deep geothermal reservoir would be located beneath the zone of intensive hydrothermal demagnetisation.

Hydrothermal activity can also produce, on a smaller scale, a '"re-magnetisation" when pyrrhotite is deposited, which occurs as a secondary mineral in some New Zealand geothermal reservoirs. Little is known about the formation of this hydrothermal, magnetic trace mineral, which often can be found near rocks containing organic matter. Its magnetic stability range is restricted by its low Curie temperature (about 320°C according to Butler [6]). A study by Browne and Ellis [5] showed that small amounts of pyrrhotite occur in about one third of the wells at Ohaaki geothermal field in the TVZ. Pyrrhotite has also been found in a few wells at two other TVZ geothermal fields, the Wairakei and Waiotapu geothermal fields [33]; it is usually confined to barren fissures, not dispersed. Some 1990s studies by students at the Geothermal Institute (University of Auckland) of cores from 11 wells at Ohaaki geothermal field indicate that the magnetic effect of this alteration mineral (if present) would be small. However, two cores from one well at Ohaaki field at about 800 and 1100 m depths, which showed up with significant magnetisation contain no magnetite but some pyrrhotite. Overall, however, it appeared that "re-magnetisation"' of reservoir rocks caused by pyrrhotite deposition is small and can be neglected.

### 3.4. Reversely magnetised volcanic rocks

Reversely magnetised volcanic rocks are rocks which have $\mathbf{M}_r$ pointing opposite to the direction of the present day earth magnetic field. These volcanic rocks have thermo remanent magnetisation (TRM) which opposed the present day geomagnetic field because they were deposited and cooled down to temperature below the Curie point of magnetite (about 580°C) the time of geomagnetic reversal. As the last reversal of the earth magnetic field occurred about

0.7 Myrs ago (Section 2.1) some older Quaternary volcanic rocks would be reversely magnetised (the Quaternary age ranges from 1.6 Myrs ago to recent). In the TVZ the strength of the reversed $\mathbf{m}_r$ is always greater than that of the induced magnetisation $\mathbf{m}_i$ [28,10]. This phenomenon is probably also true in all other Quaternary volcanic zones elsewhere. As the results, the total magnetisation ($\mathbf{m}_t$), which is the rock parameter affecting airborne magnetic anomalies, of the reversely magnetised Quaternary volcanic rocks is opposite in direction to $\mathbf{B}_0$. Reversely magnetised rocks have *negative* magnetisation contrast, the same type of magnetisation contrast as that caused by hydrothermal demagnetisation.

# 4. The pattern of total force magnetic anomalies caused by hydrothermal demagnetisation in different geographic locations

## 4.1. Pattern of the total force magnetic anomalies

To understand the magnetic anomaly pattern caused by geothermal systems at different geographical positions, the total force magnetic anomalies (ΔF) of a simplified model geothermal system (contains subsurface hydrothermally demagnetised rocks, a hill and a valley) were computed using 3D quantitative magnetic modelling code written by Soengkono [21] based on the equations of Barnett [2]. The computations were made for the geothermal system sited at different geographical locations; at high, moderate and low magnetic inclinations, at the magnetic poles and along the magnetic equator. The simplified model of geothermal system and its total force magnetic anomalies at the various magnetic inclinations are shown in Figures 4, 5, 6 and 7.

As already mentioned in Section 3.3, the hydrothermally demagnetised rocks have a *negative magnetisation contrast* with respect to the surrounding volcanic rocks. It should be noted that the magnetic model in Figures 4(a), 5(a), 6(a) and 7(a) will be still exactly the same magnetically if the hydrothermally demagnetised rocks body is given a $|\mathbf{M}_t|$ of -1.7 a/m and the surrounding homogeneous volcanic rocks mass is given a magnetisation $|\mathbf{M}_t|$ of 0 A/m. This is true because the magnetic effect of a one-dimensional magnetic body (a horizontal magnetic plate, or slab) is everywhere equal zero.

### 4.1.1. *In the high magnetic inclination*

Figure 4 shows the total force anomalies created by the simplified magnetic model at the southern hemisphere at geomagnetic inclination of -60° (Figure 4(b)) and at northern hemisphere at geomagnetic inclination of +60° (Figure 4(c)). In both Figures 4(b) and 4(c), the negative magnetisation contrast of the hydrothermally demagnetised rocks (the geothermal reservoir) causes bipolar total force magnetic anomaly with a dominant *magnetic low* (negative anomalies) that are easily recognised on the maps. Although the centre of the magnetic low is shifted from the entre of demagnetised rocks, the magnetic low can still clearly identify the demagnetised body. The extent of the magnetic low can be used almost directly to approximate (roughly delineate) the edges of the demagnetised rocks.

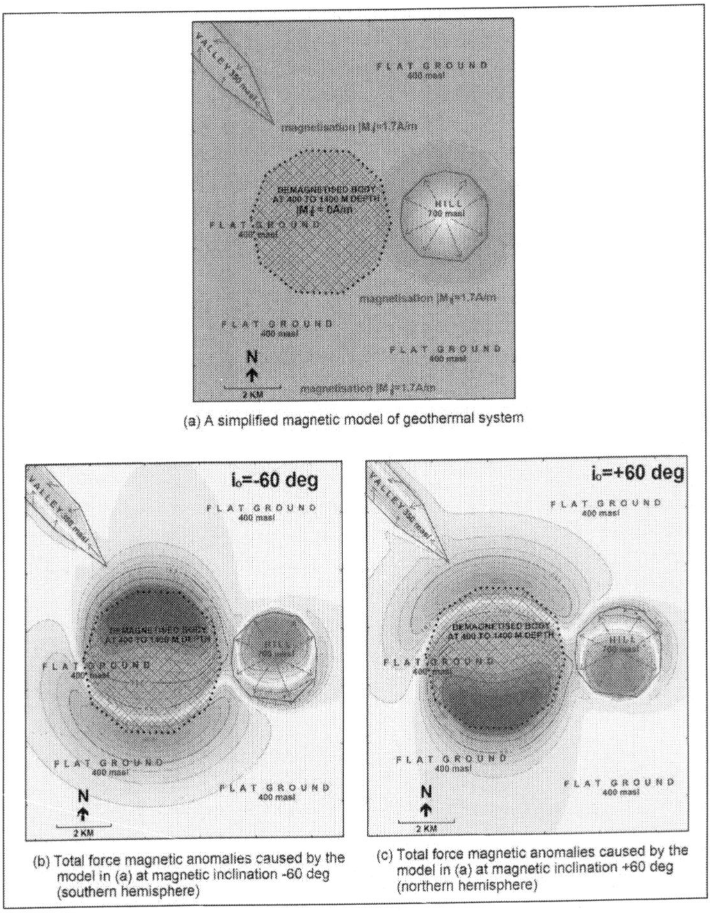


(a) A simplified magnetic model of geothermal system

(b) Total force magnetic anomalies caused by the model in (a) at magnetic inclination -60 deg (southern hemisphere)

(c) Total force magnetic anomalies caused by the model in (a) at magnetic inclination +60 deg (northern hemisphere)

**Figure 4.** Total force magnetic anomalies (ΔF) caused by a simplified geothermal system at high geomagnetic inclinations of -60° (southern hemisphere) and +60° (northern hemisphere).

#### *4.1.2. In the moderate magnetic inclination*

Figure 5 shows the total force anomalies created by the simplified magnetic model at southern hemisphere at geomagnetic inclination of -30° (Figure 5(b) and at northern hemisphere at geomagnetic inclination of +30° (Figure 5(c)). The bipolarity of the total force magnetic anomalies becomes more pronounced and the strengths (magnitudes) of the positive and negative anomalies are roughly equal. The hydrothermally demagnetised rocks can still be identified, but it becomes difficult to delineate their extent directly from the extent of the magnetic anomalies.

http://dx.doi.org/10.5772/61651

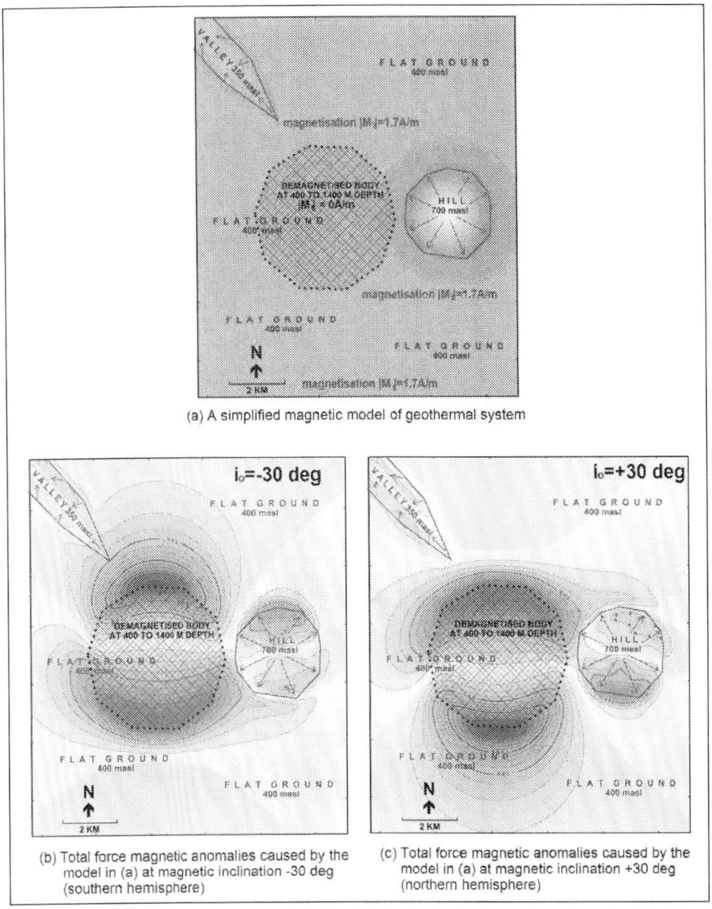


(a) A simplified magnetic model of geothermal system

(b) Total force magnetic anomalies caused by the model in (a) at magnetic inclination -30 deg (southern hemisphere)

(c) Total force magnetic anomalies caused by the model in (a) at magnetic inclination +30 deg (northern hemisphere)

**Figure 5.** Total force magnetic anomalies ($\Delta F$) caused by a simplified geothermal system at moderate geomagnetic inclinations of -30° (southern hemisphere) and +30° (northern hemisphere).

### *4.1.3. In the low magnetic inclination*

Figure 6 shows the total force anomalies created by the simplified magnetic model at locations close to the magnetic equator, at geomagnetic inclination -10° (Figure 6(b)) and at geomagnetic inclination +10° (Figure 6(c)). Here, the total force magnetic anomalies pattern caused by the geothermal system is much less bipolar but *positive* anomalies become dominant. Hence, it is important to be always aware to this contradictory phenomenon (*negative* magnetisation contrast creating *positive* anomalies). In this low geomagnetic latitude hydrothermally demagnetised rocks are to be identified from positive anomalies, not negative anomalies. The positive anomalies in Figures 6(a) and 6(b) are clearly recognizable, however the extent of the positive anomaly does not follow the edge of hydrothermally demagnetized rocks very well.

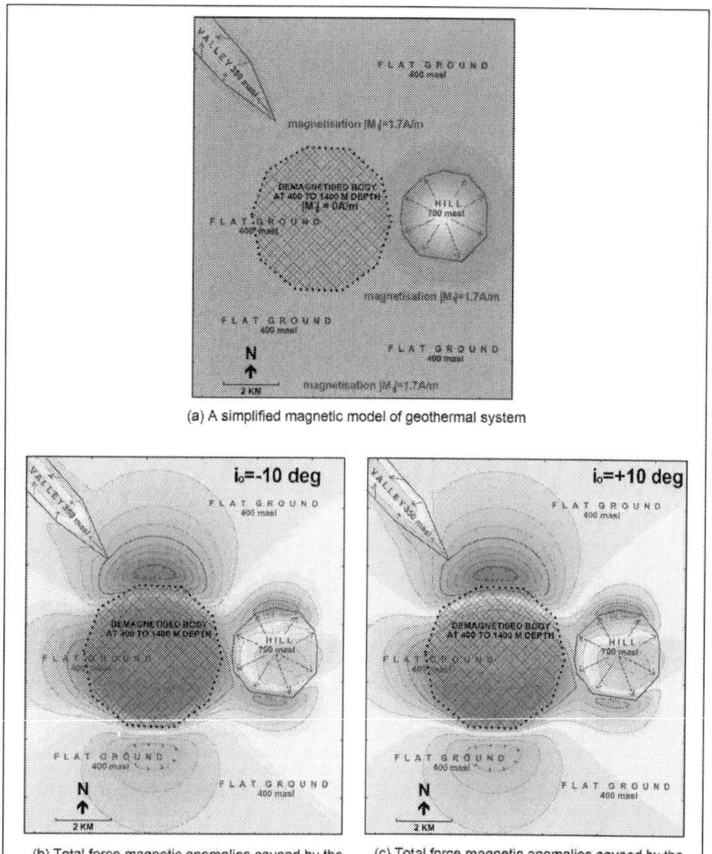


(a) A simplified magnetic model of geothermal system

(b) Total force magnetic anomalies caused by the model in (a) at magnetic inclination -10 deg (southern hemisphere)

(c) Total force magnetic anomalies caused by the model in (a) at magnetic inclination +10 deg (northern hemisphere)

**Figure 6.** Total force magnetic anomalies (ΔF) caused by a simplified geothermal system at low geomagnetic inclinations of -10° (southern hemisphere close to magnetic equator) and +10° (northern hemisphere close to magnetic equator).

### *4.1.4. Along the magnetic equator and in the south/north magnetic poles*

Figure 7 shows the total force anomalies created by the simple model at two extreme locations, along the magnetic equator where the geomagnetic inclination is 0° (Figure 7(b) and in the south/north geomagnetic pole where the geomagnetic inclination is either -90° or +90° (Figure

http://dx.doi.org/10.5772/61651

# 5. Examples of airborne magnetic surveys to investigate high temperature geothermal reservoirs

## 5.1. Introduction

A new, simple but effective, interpretation approach of airborne magnetic survey for investigation of high temperature geothermal resources in Quaternary volcanic setting has been introduced and discussed in the previous sections. To gain a more comprehension of its practical aspects, the interpretation approach is applied to real airborne magnetic data from the TVZ in New Zealand (geomagnetic inclination about -65°) and from the eastern Java Island in Indonesia (geomagnetic inclination about -35°). These examples of interpretations are presented and discussed in the following sections. Note that all the magnetic interpretations presented are new interpretations, specifically carried out to illustrate the approach introduced previously in this chapter. The gridding, plotting contouring and drawing, trend determinations and RTP transformations of the anomalies are all carried out using the Oasis Montaj software package.

## 5.2. Examples from TVZ, New Zealand

### *5.2.1. Wairakei geothermal field*

The high temperature Wairakei geothermal field in the TVZ is the first geothermal field used for electricity generation in New Zealand, and the second in the world after the Larderello geothermal field in Italy. The geothermal system is situated in rather flat topography (Figure 10). The geothermal reservoir was delineated using Schlumberger DC resistivity surveys and the boundary shown in Figure 10 has been slightly refined using information from a few geothermal boreholes. To the southeast of Wairakei is another high temperature field, the Tauhara geothermal field. Both Wairakei and Tauhara reservoir are hosted by Quaternary volcanic rocks. The fields are located at a high magnetic inclination of about -65°. Figure 11 shows the map of $|\mathbf{B}_{obs}|$ over the Wairakei field and the northern part of the Tauhara field obtained from a detailed airborne magnetic survey draped 60 m above ground which was conducted by the gold exploration company Glass Earth NZ Ltd in 2006, and from a previous smaller survey at similar altitude conducted by the GNS Science in 1989. The 1st order trend of $|\mathbf{B}_{obs}|$ to estimate $|\mathbf{B}_o|$ is also shown in Figure 11. Figure 12 presents the total force magnetic anomalies ($\Delta F$) map over the area.

The total force magnetic anomalies after the reduction to pole (RTP) transformation are shown in Figure 13. The magnetic anomalies (RTP) shown in this figure are located directly above their causative sources. Prominent magnetic lows are present in the north-western part of the Wairakei field and over the Tauhara field. These magnetic lows represent intensive and/or thick hydrothermal demagnetisation. Less prominent magnetic low can be seen over the fumaroles and steaming ground at Crater of the Moon, representing hydrothermal demagnetisation by acidic condensate in the shallow vapour zones. The prominent magnetic low in the north-western part of the Wairakei field suggests that in this area the Wairakei geothermal reservoir extends beyond the field boundary defined by resistivity survey, although the possibility that part of the negative anomalies outside the boundary of the geothermal field are due to by reversely magnetised rocks cannot be completely ruled out.

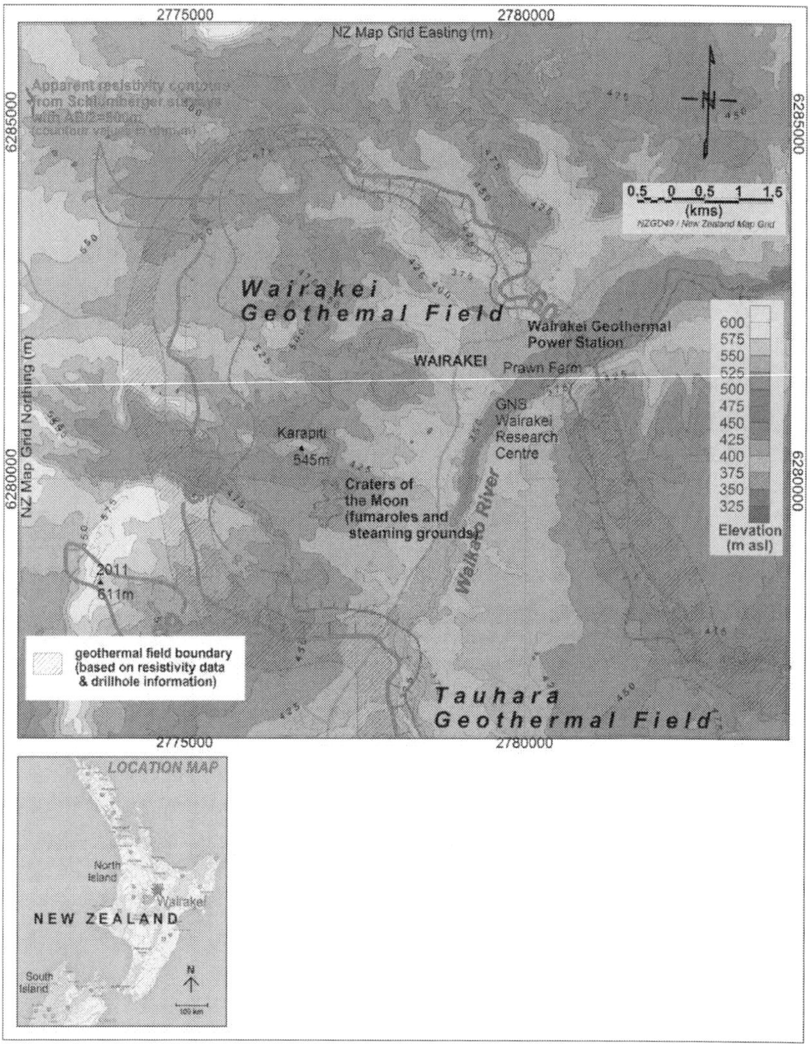


**Figure 10.** Topographic map of the Wairakei and the northern part of Tauhara geothermal fields in the TVZ (New Zealand) showing the boundary of the fields based on Schlumberger DC resistivity survey.

http://dx.doi.org/10.5772/61651

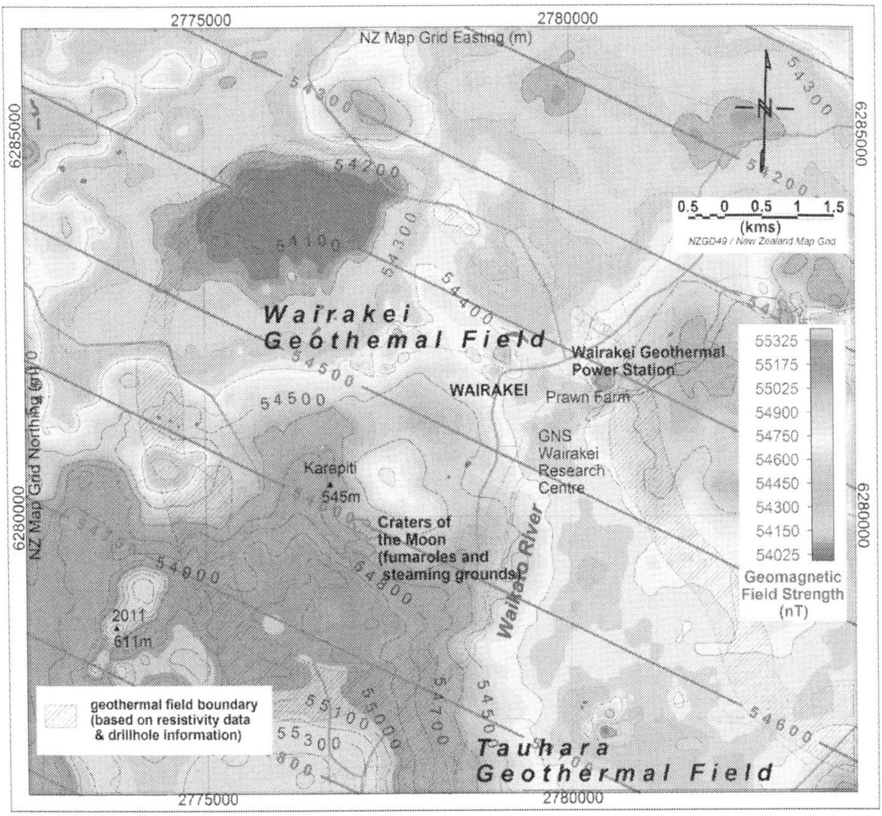


**Figure 11.** Map of observed total force geomagnetic field strengths ($|\mathbf{B}_0|$) over the Wairakei and the northern part of Tauhara geothermal fields in the TVZ (New Zealand). The red contour lines represent $|\mathbf{B}_0|$ values as defined by the first order trend of all $|\mathbf{B}_{obs}|$ across the area.

### *5.2.2. Mokai geothermal field*

The high temperature Mokai geothermal field is also located in the Taupo Volcanic Zone, about 20 km west of the Wairakei field. The Mokai geothermal reservoir is also hosted by Quaternary volcanic rocks and has slightly steeper topography than the Wairakei field (Figure 14).

Figure 14 shows the boundary of Mokai geothermal field delineated from Schlumberger DC resistivity surveys, and the surface thermal manifestations of the field. The 60 ohm-m resistivity contour is extending northeast-north toward the thermal springs in a lower elevation, indicating lateral outflow of geothermal water. Figure 15 shows the map of $|\mathbf{B}_{obs}|$ over the Mokai geothermal field obtained from the same detailed survey by Glass Earth Ltd. in 2006

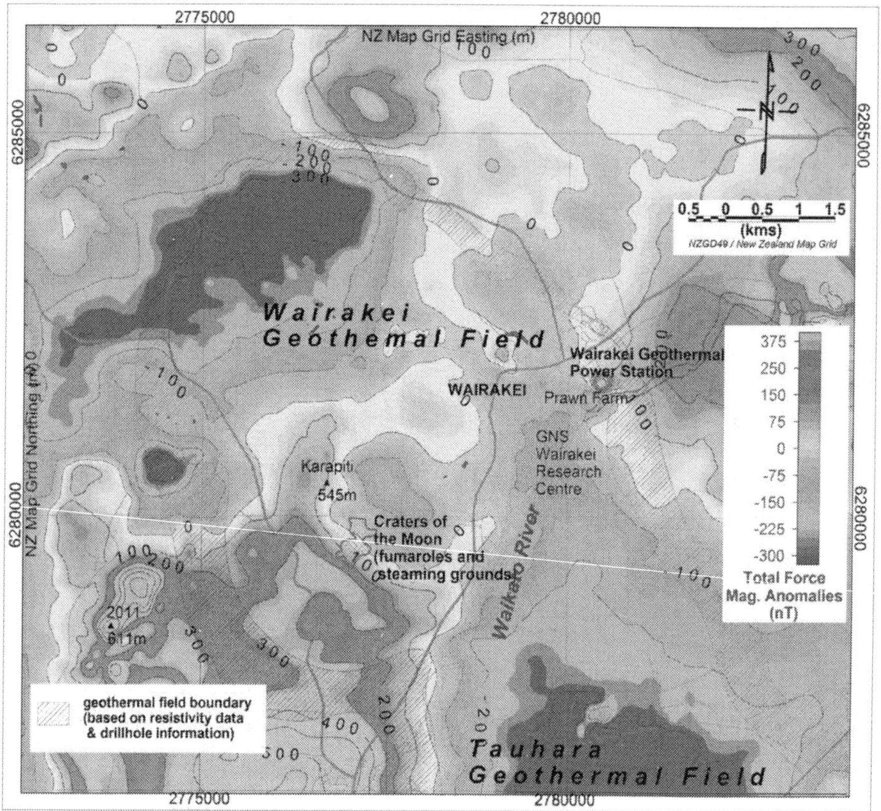


**Figure 12.** Map of total force magnetic anomalies (ΔF) over the Wairakei and the northern part of Tauhara geothermal fields in the TVZ (New Zealand). The ΔF values were obtained by subtraction of $|\mathbf{B}_0|$ determined from 1st order trend of $|\mathbf{B}_{obs}|$.

that covers the Wairakei field. The 1st order trend of $|\mathbf{B}_{obs}|$ to estimate $|\mathbf{B}_0|$ is also shown in the map. The total force magnetic anomalies (ΔF) map the Mokai area is shown in Figure 16. A broad magnetic low appears associated with the geothermal field at location slightly shifted to the north. Several other magnetic lows are present outside the Mokai geothermal field to the northwest, southwest, southeast and further to the northeast of the Mokai resistivity boundary.

Reduction to pole (RTP) transformation was applied to the total force anomalies in Figure 16 and the result is presented in Figure 17. In this figure, the magnetic low above the Mokai geothermal field becomes consistent with the resistivity boundary, indicating that hydrother-

http://dx.doi.org/10.5772/61651

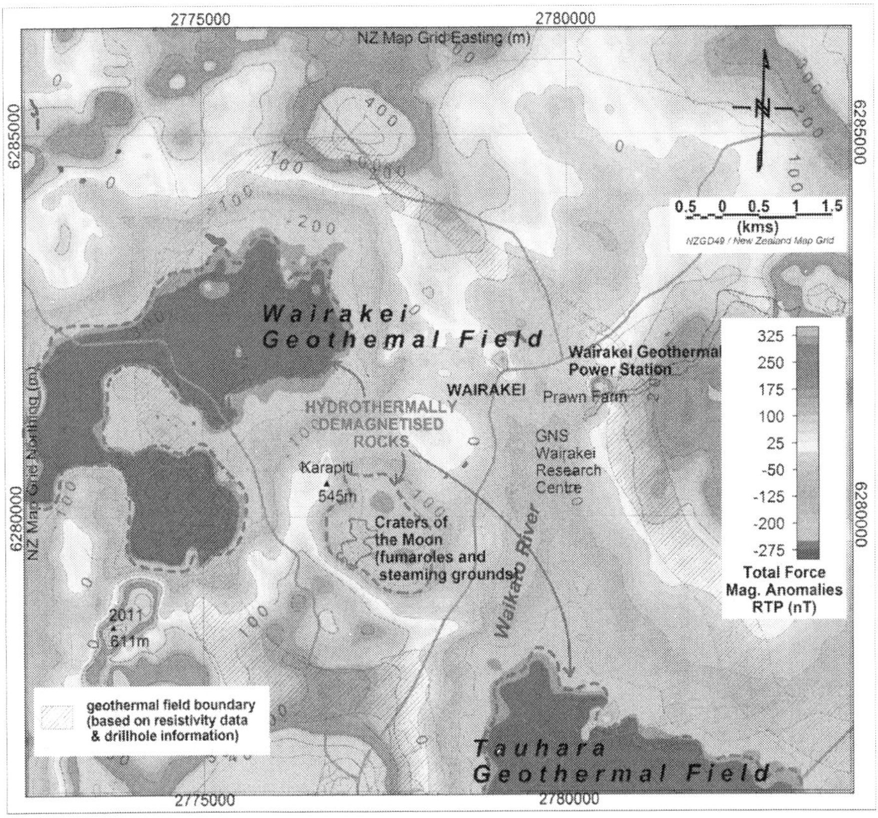


**Figure 13.** The result of reduction to pole (RTP) of the total force magnetic anomalies (ΔF) over the Wairakei and the northern part of Tauhara geothermal fields in the TVZ (New Zealand).

mally demagnetised rocks are present in the Mokai Reservoir. Moderately low magnetic RTP anomalies mark the western part of Pukemoremoe rhyolite topographic dome, suggesting that acidic condensate formed in shallow vapour zones here has caused hydrothermal demagnetisation. Reversely magnetised rocks are known to be present in this area [28]. Hence, the magnetic lows outside the Mokai geothermal field boundary could represent reversely magnetised rocks, as indicated in Figure 16. The magnetic low caused by the hydrothermally demagnetised rocks inside the Mokai reservoir shows variation in strength. This variation could be caused the variation of intensity and/or thickness of the hydrothermal demagnetisation process, which could held clue to the variation of reservoir permeability and/or movement of geothermal water. This could be investigated further by a quantitative 3D modelling of the magnetic anomalies.

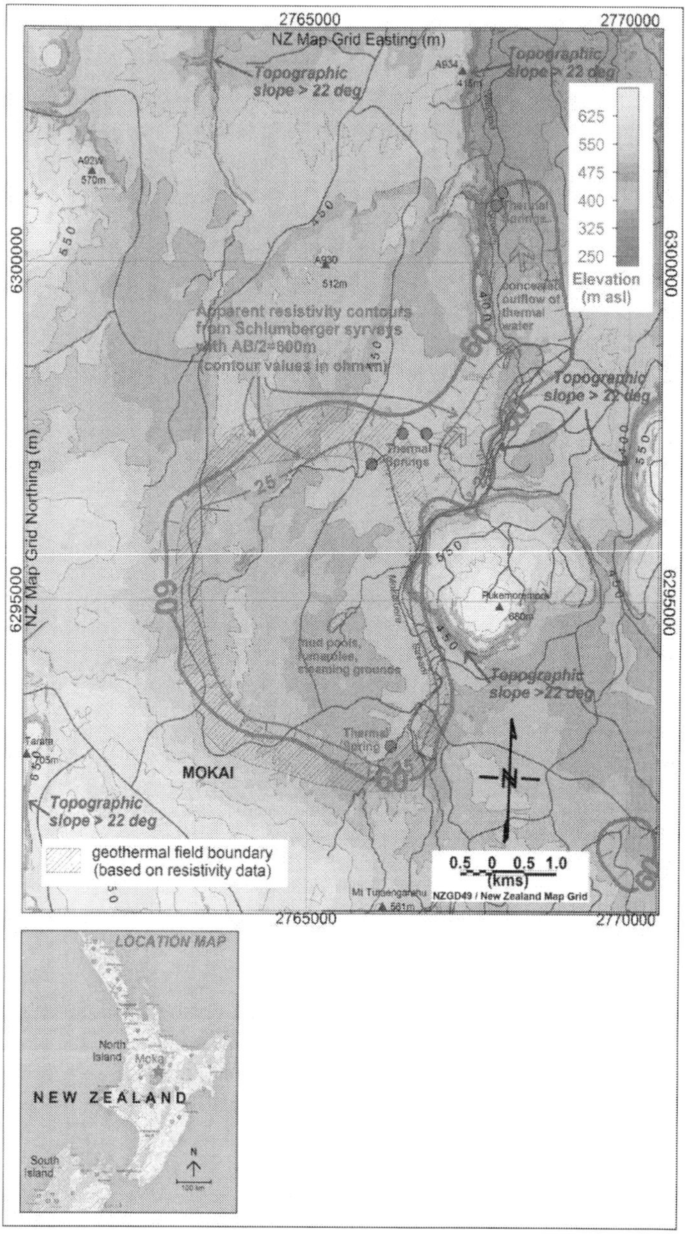


**Figure 14.** Topographic map of the Mokai geothermal field in the TVZ (New Zealand) showing the boundary of the field based on Schlumberger DC resistivity survey.

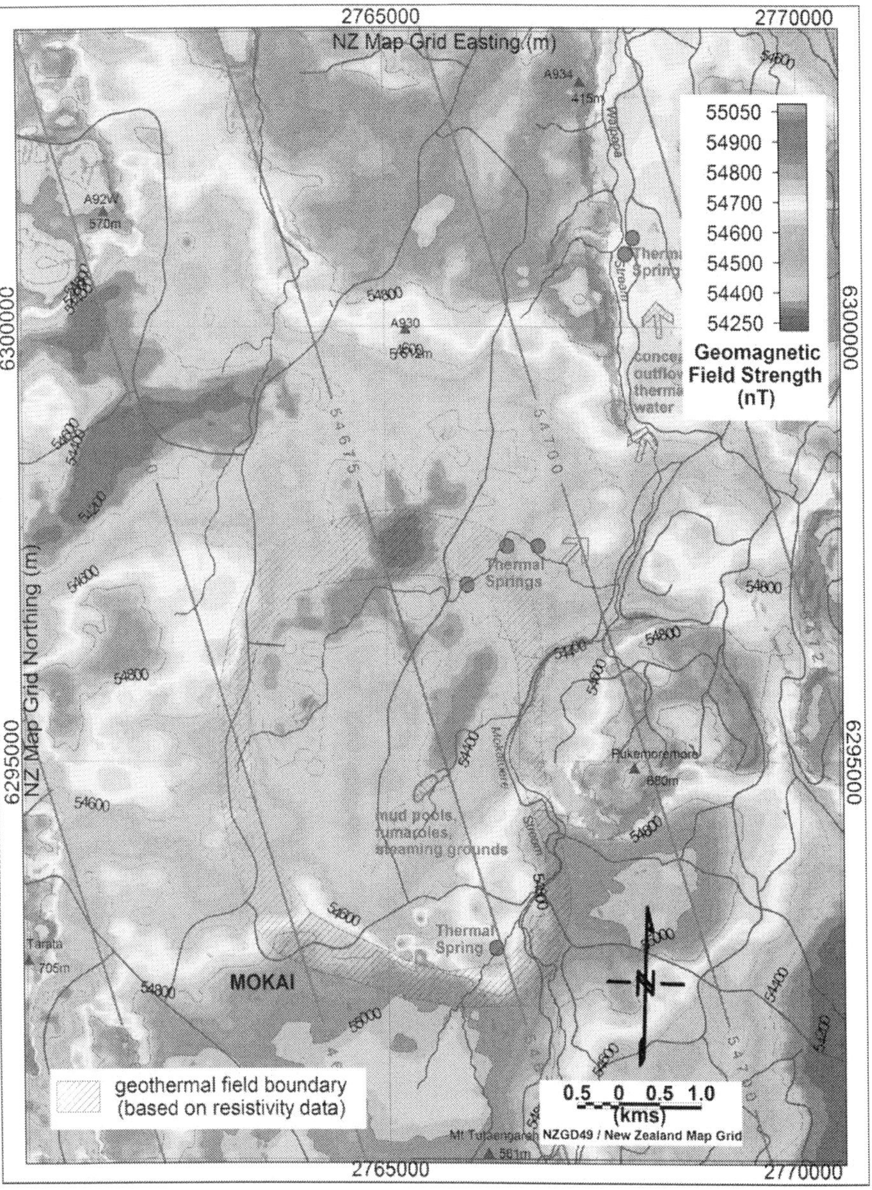


**Figure 15.** Map of observed total force geomagnetic field strengths (|$\mathbf{B}_0$|) over the Mokai geothermal field in the TVZ (New Zealand). The red contour lines represent |$\mathbf{B}_0$| values as defined by the first order trend of all |$\mathbf{B}_{obs}$| across the area.

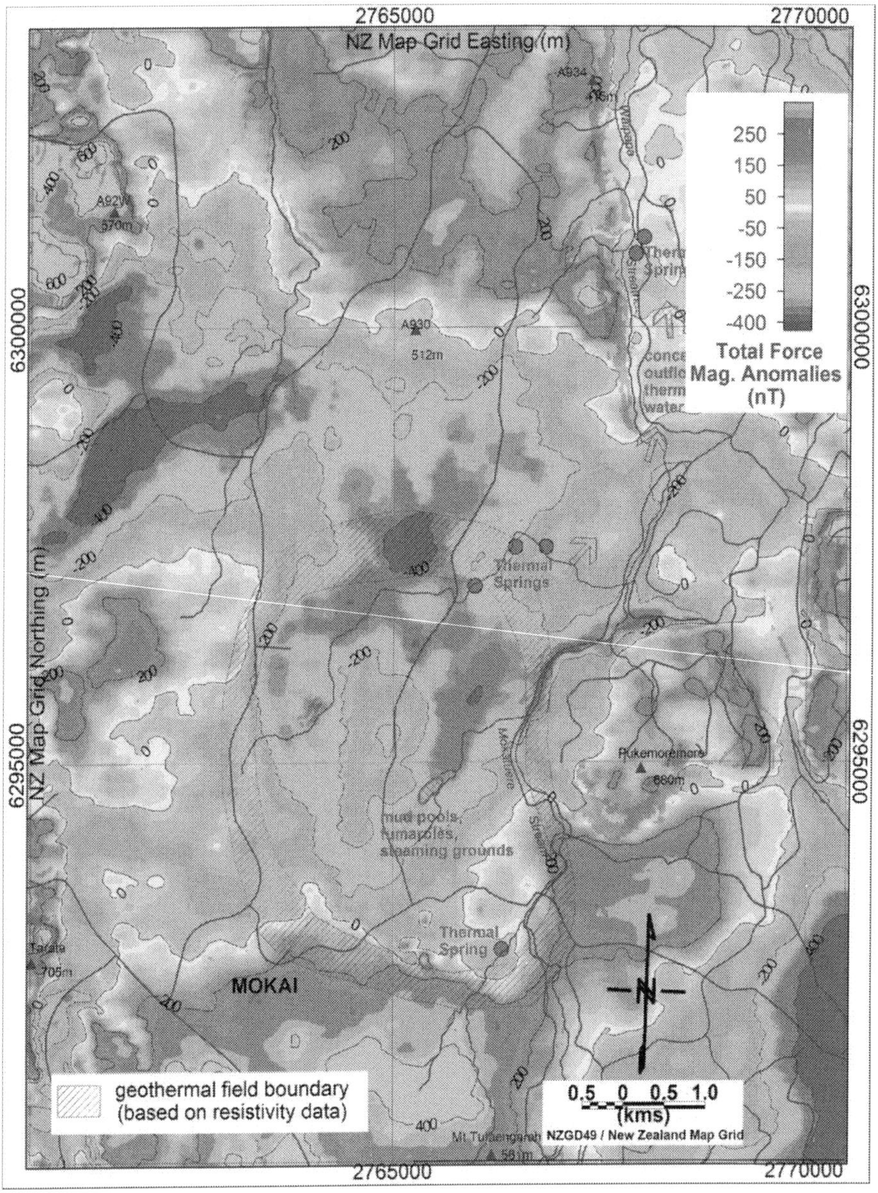


**Figure 16.** Map of total force magnetic anomalies ($\Delta F$) over the Mokai geothermal field in the TVZ (New Zealand). The $\Delta F$ values were obtained by subtraction of $|\mathbf{B}_0|$ determined from 1st order trend of $|\mathbf{B}_{obs}|$.

http://dx.doi.org/10.5772/61651

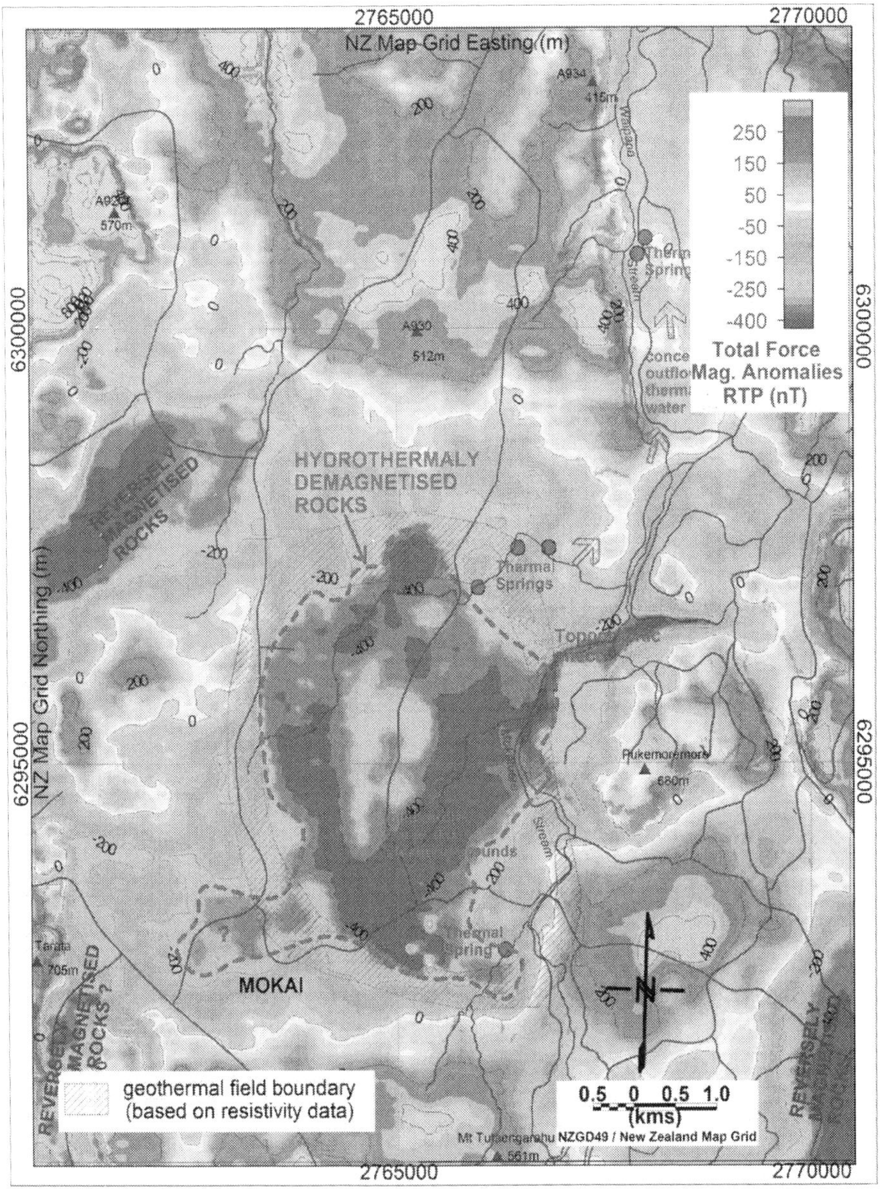


**Figure 17.** The result of reduction to pole (RTP) of the total force magnetic anomalies (ΔF) over the Mokai geothermal field in the TVZ (New Zealand).

### 5.3. Example from Java Island, Indonesia

#### *5.3.1. Ijen geothermal field*

The Ijen geothermal field is located in eastern Java Island of Indonesia (see the location map in Figure 18) with a geomagnetic inclination of about -35°. The airborne magnetic survey over the Ijen geothermal field was carried out in 1990 by Penas-Carson Services Inc. (USA) for the Indonesia Pertamina Geothermal Division. The survey was made at flight elevation of about 1 km above the ground, along west-east flight lines separated by about 0.75-1 km spacing. The Ijen geothermal field is hosted by Quaternary andesitic volcanic rocks which form steep topography around the field. Figure 18 shows the topography of the Ijen area and the locations of surface thermal manifestation of the geothermal field consisting of altered rocks near Telagawaru and a group of thermal springs near Blawan in the north. The "raw" airborne magnetic data were presented by the Penas-Carson Inc., which were obtained by reducing the IGRF variation from the measured geomagnetic field. Such "raw" airborne magnetic data are shown in Figure 19, together with their 1$^{st}$ order trend. Figure 20 shows the total force magnetic anomalies ($\Delta F$) over the Ijen area. In this figure the bipolar anomalies associated with Mt Suket, Mt Pendil and Mt Rante become clearly visible. Less clearly shown is the bipolar anomaly over Mt Ijen. Positive anomalies are also seen over the north-eastern Kendeng Ridge. A wide magnetic low is presents over the geothermal field region. The north-eastern part of this magnetic low could be the negative part of the bipolar magnetic anomaly of the north east Kendeng Ridge. Because of the moderate magnetic inclination of the region, the $\Delta F$ values shown in Figure 20 are likely to spread widely over the causative sources. A direct interpretation of this figure can be misleading without the RTP transformation.

The total force anomalies over the Ijen geothermal area after reduction to pole (RTP) transformation are shown in Figure 21. The magnetic anomalies (RTP) shown in this figure would be located above their causative sources. The most interesting magnetic RTP anomalies in Figure 21 is the magnetic low that appears to be associated with the outcrop of altered rocks west of Telagawaru. This magnetic low is interpreted in Figure 21 to represent hydrothermally demagnetised rocks. This leads to the inferred model of Ijen geothermal system consisting of a geothermal up-low zone in the Telagawaru – Mt Genteng area and concealed outflow of thermal water towards the thermal springs near Blawan. The magnetic low to the northwest of Kendeng Ridge could represent reversely magnetised Quaternary volcanic rocks. Reversely magnetised rocks are probably also the sources of the three magnetic lows near the southern edge of Figure 21. Positive anomalies marked Mt Suket, Mt Pendil, Mt Rante and Mt Ijen, showing that these mountains are formed by normally magnetised rocks. The magnetic high which follows the north-eastern Kendeng Ridge in Figure 20 has moved in Figure 21 to a new location to be entirely southwest of the ridge. This magnetic high probably represents a subsurface lava body underneath.

http://dx.doi.org/10.5772/61651

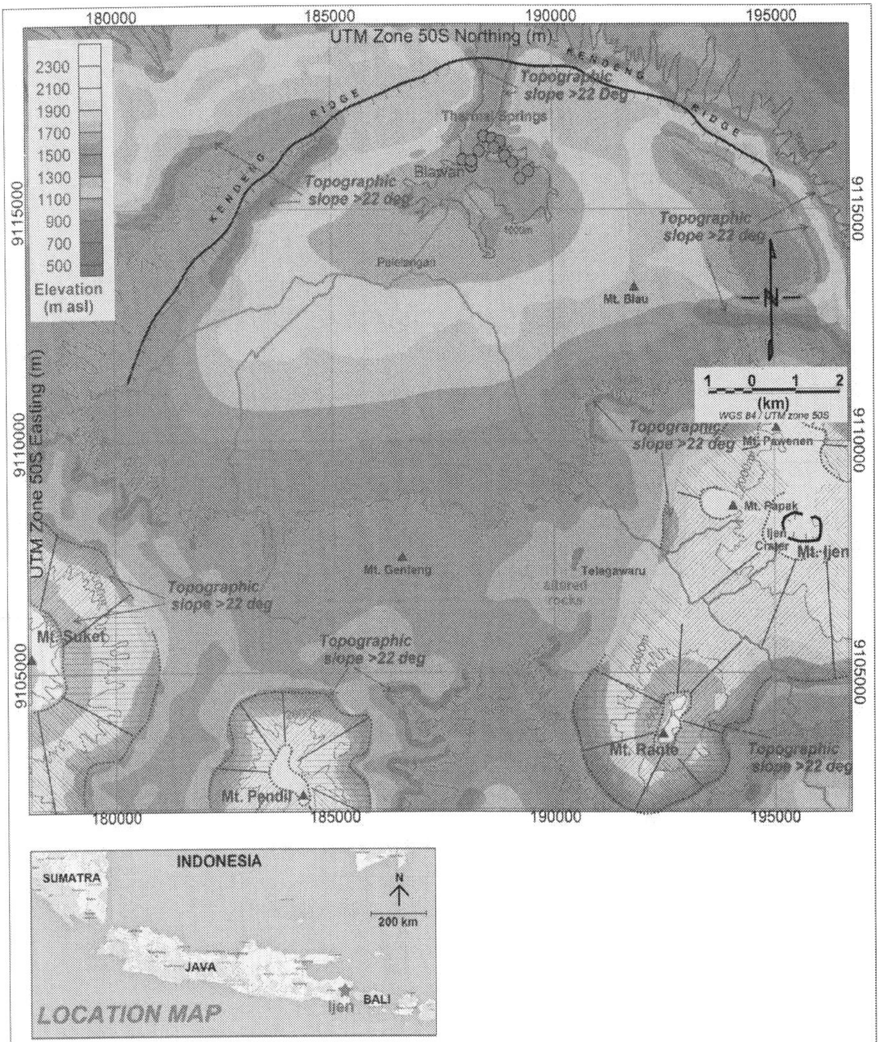


**Figure 18.** Topographic map of the Ijen geothermal field in the Eastern Java (Indonesia) showing the prominent topography of the surrounding Mt. Suket, Mt. Pendil, Mt Rante and Mt Ijen and the surface thermal manifestations (altered rocks and thermal springs) of the field.

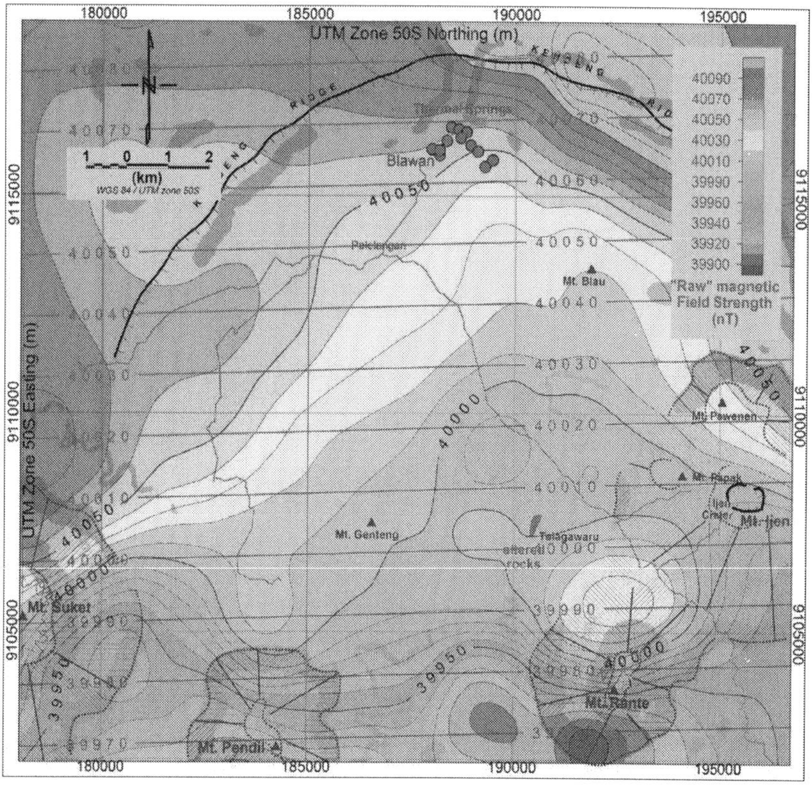


**Figure 19.** Map of observed total force "raw" field strengths (~$|\mathbf{B}_{obs}|$) over the Ijen geothermal field in the eastern Java (Indonesia). The red contour lines represent $|\mathbf{B}_{o}|$ values as defined by the first order trend of all ~$|\mathbf{B}_{obs}|$ across the area.

# 6. Summary and discussion

This chapter has shown that airborne magnetic data can be very useful in the investigation of high temperature geothermal reservoirs hosted by Quaternary volcanic rocks, particularly in the area with difficult ground access. Data might be already available over the geothermal target area from some previous surveys by mineral exploration companies or government institutions. To carry out a new airborne magnetic survey, many geophysical exploration companies are advertising their service and can be contacted on the internet, some of them can do the survey almost anywhere in the world. Even when a new survey is required, the airborne magnetic survey should still be a cost-effective method to explore and investigate high temperature geothermal resources in Quaternary volcanic setting.

# Chapter 6

# Geothermal Exploration Methods

Essam Aboud

Additional information is available at the end of the chapter

http://dx.doi.org/10.5772/61679

**Abstract**

A geothermal reservoir causes inhomogeneity in the physical properties of the subsurface geology due to the high changes in temperature. These physical properties can be observed by means of varying anomalies from geophysical observations from the surface. These physical properties include electrical conductivity, rock density, magnetic susceptibility, rock elasticity, and, finally, the temperature. The above mentioned physical properties can be detected by surficial geophysical survey. The same parameters can also be measured in wells using "geophysical well logging," providing data that are more accurate but costly. On the other hand, geochemical exploration assists in gathering information about the subsurface composition of the fluids. This information can be used indirectly to know the temperature, origin, and flow direction, which help in locating subsurface geothermal reservoir. It is clear that, from geophysical and geochemical methods, shape, size, structure, depth, and heat sources of the reservoir can be traced and mapped. Thus, the geophysical and geochemical surveys play a key role in geothermal exploration. This chapter will discuss the above mentioned methods in detail presenting some examples from literature review.

**Keywords:** Geothermal exploration, Geophysics, Geochemistry

## 1. Introduction

Geothermal energy is "heat" contained in the earth's interior. This heat comes from the earth's core continuously outward until it traps in impermeable and fractured layers of the earth's surface. When water is heated by this "heat," hot water or steam can be trapped in permeable and porous rocks under a layer of impermeable rock, forming *geothermal reservoir* or *geothermal system*. Geothermal reservoir can be described schematically as convective water in the upper crust of the earth, which transfers the heat from a heat source to a heat sink [4]. A geothermal reservoir is, in general, composed of three main elements: a *heat source*, a *reservoir*, and a *fluid*, the carrier that transfers the heat. The *heat source* can be either a very high-temperature (> 600

°C) magmatic intrusion, which has reached relatively shallow depths (5–10 km), or low-temperature zones (e.g., volcanic zones). The *reservoir* is a volume of hot permeable rocks through which the circulating fluids extract heat. It is often overlain by a cover of impermeable rocks and connected to a surficial recharge area through which the meteoric waters can replace, or partly replace, the fluids that run from the reservoir through springs or are extracted by boreholes. The *fluid or geothermal fluid* is water, meteoric water, in the liquid or vapor phase, depending on its temperature and pressure. This water often carries with it chemicals and gases such as $CO_2$ and $H_2S$. Figure (1) is a simplified representation of an ideal geothermal reservoir.

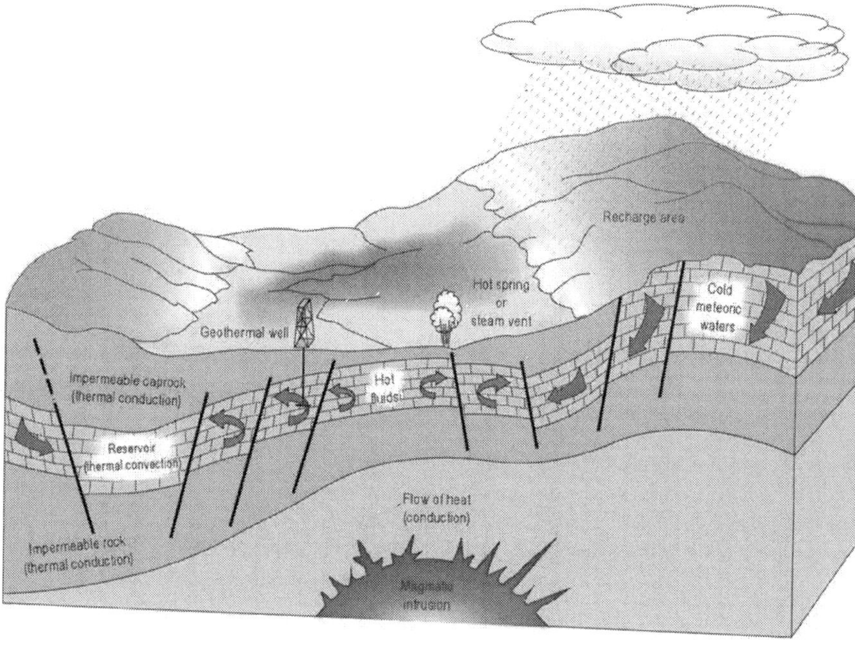


**Figure 1.** Schematic diagram of an ideal geothermal system by International Geothermal Association (IGA).

A geothermal reservoir in general causes inhomogeneity in the physical properties of the subsurface geology (e.g., physical properties of rocks) due to the high changes in temperature. These physical properties can be observed by means of varying anomalies from geophysical observations from the surface. These physical properties include electrical conductivity (electrical/electromagnetic (EM) method), rock density (gravity method), magnetic susceptibility (magnetic method), rock elasticity (seismic method), and, finally, the temperature (thermal survey). The above mentioned physical properties can be detected by surficial geophysical survey. The same parameters can also be measured in wells using "geophysical well logging," providing data that are more accurate but costly. On the other hand, geochemical exploration assists in gathering information about the subsurface composition of the fluids. This information can be used indirectly to know the temperature, origin, and flow direction, which help in locating subsurface geothermal reservoir. It is clear that, from geophysical and geochemical methods, shape, size, structure, depth, and heat sources of the reservoir can be traced and mapped. Thus, the geophysical and geochemical surveys play a key role in geothermal exploration. This chapter will discuss the above mentioned methods in detail, presenting some examples from literature review.

Geological, geophysical, and geochemical data must be collected and integrated with any available data from previous studies on water, minerals, etc. in the study area and adjacent areas before geothermal exploration program. This information plays an important role in defining the objectives of the geothermal exploration program and could lead to a significant reduction in costs. The classical geothermal exploration program includes the following phases:

1. Reconnaissance field survey,
2. Geochemical surveys,
3. Geophysical surveys,
4. Exploratory well, and
5. Conceptual model.

## 2. Reconnaissance field survey

Reconnaissance field survey or preliminary survey in geothermal exploration program is considered as the initiation phase on which next phases are defined. This phase involves a work program to assess the available evidence for geothermal potential in a specific area. While some of these evidences are technical (e.g., geological data, thermal manifestation, etc.), others are social (e.g., land access, country regulations, logistics, etc.). The preliminary survey seeks to identify geological settings that might host economically geothermal reservoirs. In practice, the survey essentially involves an "office work" review of geological, hydrological, and/or hot spring/thermal data, drilling data, remote sensing, etc.

The reconnaissance survey phase should also include an assessment of key environmental issues or factors that might affect or be affected by a geothermal project. As with any major infrastructure development, geothermal power plants have their own unique social and environmental impacts and risks that require awareness and management. All these factors can significantly affect the time and cost required to move through the subsequent phases of project. The reconnaissance survey aims to show whether the area of interest has a geological setting or features that may indicate the presence of an economically exploitable geothermal system. Once this is established, the project developer must then determine the feasibility of obtaining concessions over the most promising areas and, if they become productive, how would geothermal power fit with the existing energy infrastructure. Although the reconnaissance survey phase is primarily "desk based," one or more short field visits might greatly assist in confirming the geothermal play type(s), the regional geology, the surface thermal features, and in identifying key environmental and social issues. In general, the basic information collected during the reconnaissance survey phase covers:

- The power market and possible power purchase agreements or feed-in tariff.
- Other/additional demands and possibilities for geothermal energy use such as district or greenhouse heating.
- Infrastructure issues (roads, water, communication, transmission).
- Resource ownership issues (in some countries, geothermal permits are under mining laws; elsewhere, it may be considered a water right under specific geothermal legislation; or a relevant legal framework might not yet exist).
- Environmental and social issues.
- Institutional and regulatory frameworks.
- Issues relating to political and financial stability.
- Collection and interpretation of available remote sensing or aerial survey data.
- Information from available literature on any known geothermal systems, including geological, hydrological, and/or hot spring/thermal data and historic exploration data.
- Information from previous explorations or wells that may have been drilled in the area of interest.

Based on the outcomes of the reconnaissance survey, the explorer or developer may decide to proceed to the next step: "exploration phase." Obtaining finance and/or partners to share the risks and expenses of this phase may also be necessary. There may be several potential sites to investigate, which could effectively spread the risk but require higher overall expenditures. Engaging experienced geothermal consultants during the reconnaissance survey phase is one of the keys to identifying and thoroughly assessing relevant background information, identifying possible non-geological issues, and designing an effective forward exploration program. The time required for the reconnaissance survey phase depends on a range of factors. The time may be as short as several months. However, if there are many potential sites to investigate and if environmental approvals and the permit process are complex and finance is difficult to secure, the survey may take a year or longer.

## 3. Geochemical surveys

Geochemical surveys are used to determine whether the geothermal reservoir is water or vapor dominated, estimate the minimum temperature expected at depth, estimate the homogeneity

http://dx.doi.org/10.5772/61679

of the water supply, infer the chemical characteristics of the deep fluid, and determine the source of recharge water. The geochemical survey consists of sampling and chemical and/or isotope analyses of the water and gas from geothermal manifestations (e.g., hot springs, fumaroles) or wells. This survey provides useful data for planning exploration, and its cost is relatively low compared to other more sophisticated methods, such as the geophysical surveys. Geochemical surveys with the use of tracers can also offer information on the direction of movement of subsurface groundwater and of reinjected fluids.

### 3.1. Field and laboratory analysis

From the reconnaissance field, comparison between the shape and type of topography, and the thermal emissions that existed can roughly suggest the presence of a high enthalpy system at depth. Then, the chemical composition analysis can definitely suggest whether that area deserves to be further investigated using geophysical methods and eventually through deep drilling. In general, geochemical analysis/application (e.g., geothermometers) in liquid and gas phase gives reliable estimates of temperature of the sources of fluids, whose depth cannot generally be derived by the geochemical prospecting.

For the above reasons, fieldworks for sampling includes thermal water and gas samples, associated either with the springs or as dry emissions, as well as hydrothermal deposits around the sites of investigations. For example, in silicic formations, thermal springs, at the surface, precipitate silica and Fe hydroxides. In some cases, travertine precipitates from thermal springs [7,2], such as those areas located along the Tethys orogenic belt giving rise to huge deposits, and sometimes 100 m thick. During spring's sampling, temperature, pH, electrical conductivity, as well as the concentration of $HCO_3$ must be determined in the field (special bottle must be used for silica).

In fact, as silica is an important parameter in geothermal exploration, hot springs precipitate silica after cooling, and thermal spring water is generally stored diluted 1:10, in a separate plastic bottle in order to avoid silica precipitation during sampling. Another important factor in geothermal exploration, ammonia, should be analyzed in the field, using portable spectrophotometers. If this is not available, a fraction of the sampled water must be acidified to prevent both oxidation of ammonia and oxidation of other cations, as well as Ca, Fe, etc. precipitation as $CaCO_3$, $Fe(OH)_3$, etc.

Gas sampling is much more important than the springs. With respect to spring waters, which undergo easily dilution and mixing during underground motion, that quite often do not allow to assess the real composition of the original deep hydrothermal solutions, the gas phase rising the crust is less sensitive to dilution and mixing with shallow gases (i.e., atmospheric). Currently, sampling techniques for gases are quite well developed [3]. Air contamination should be avoided during sampling. There are a large number of components that can be measured in both liquid and gas phases. The geothermal prospector's minimum requirement in liquid/gas phase is listed in Table (1).

| | Liquid | Gas |
|---|---|---|
| Main component | Ca, Mg, Na, K, $HCO_3$, $SO_4$, Cl | $CO_2$, $N_2$, $H_2S$, $CH_4$ |
| Minor component | $SiO_2$, $NH_4$, B, Br, Sr | He, Ar, Ne, $H_2$ |
| Isotope ratio | $^{18}O/^{16}O$ and D/H | $^{13}C/^{12}C$ ratio in $CO_2$ $CH_4$ and $^{3}He/^{4}He$ ratio |

**Table 1.** Minimum requirements for liquid and gas samples

# 4. Geophysical surveys

Geophysical surveys are essential tools in geothermal exploration [12]. They allow us to detect rock and fluid properties and the existence of reservoirs and permeability pathways. Geophysical surveys (e.g., seismic, gravity, magnetic, electrical/electromagnetic, and thermal) can be defined as indirect methods. These methods, in fact, are not directly associated with the properties of the hot fluids that are being sought. Rather, they yield information about the attitude and nature of the host rocks. Detailed description of these methods will be discussed in the context.

## 4.1. Seismic survey

Seismic survey measures the "acoustic impedance" (the product of rock density and seismic velocity). Seismic survey can be divided into two subcategories based on the source of the seismic signal, artificial or natural source (commonly known as passive and active sources). Both the artificial source, such as seismic vibrator (commonly known as vibroseis), and the natural sources, such as earthquakes, volcanic eruptions, or other tectonic activity, generate seismic signal. The surveys can yield important information on the location and orientation of subsurface structures, such as faults and rock discontinuities, which may help to explain the fluid flow.

### *4.1.1. Seismic Survey Concepts*

A seismic survey is an "active" technique that images boundaries between layers of different acoustic impedance and requires a controlled source of seismic energy, such as seismic vibrators, dynamite explosives, or air guns for marine surveys. The general principle of seismic reflection is to send elastic waves from the source (e.g., seismic signal) into the underground, where each layer reflects a part of the wave's energy and allows the rest to refract through. A number of seismic receivers (geophones) that sense the motion of the ground in which they are placed record the reflected wave field at the surface (Figure 2). Surveys can be designed to image the underground along a profile (2D survey) or within a volume (3D survey).

Processed seismic data are most commonly presented as cross sections or slices (horizontal and vertical) from a seismic cube, with two-way travel time converted to true depth using the

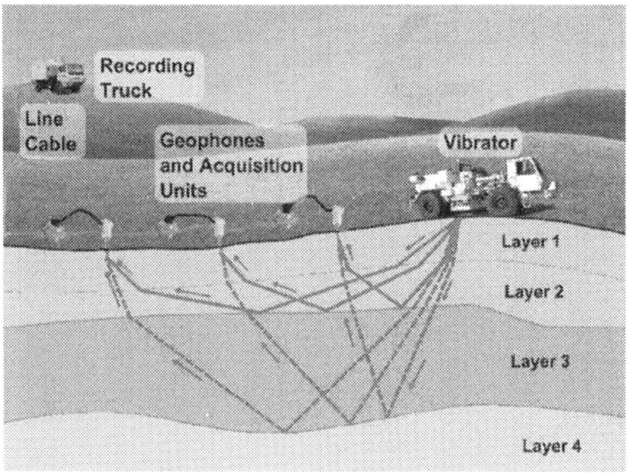


(Source: HarbourDom GmbH, Germany)

**Figure 2.** Main components of a seismic survey.

seismic velocity model and seismic migration techniques. Interpreted sections typically show the most important seismic reflectors and faults as shown in Figures (3) and (4).

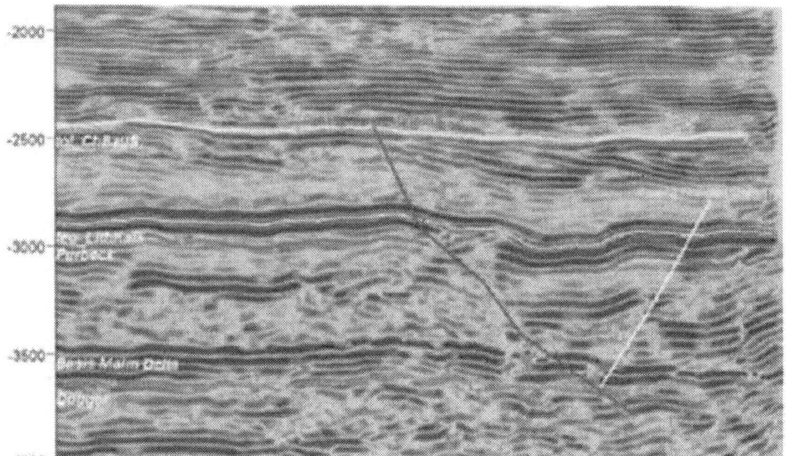


(Source: Erdwärme BayernGmbH & Co. KG)

**Figure 3.** Interpreted seismic reflection cross section with important reflectors highlighted.

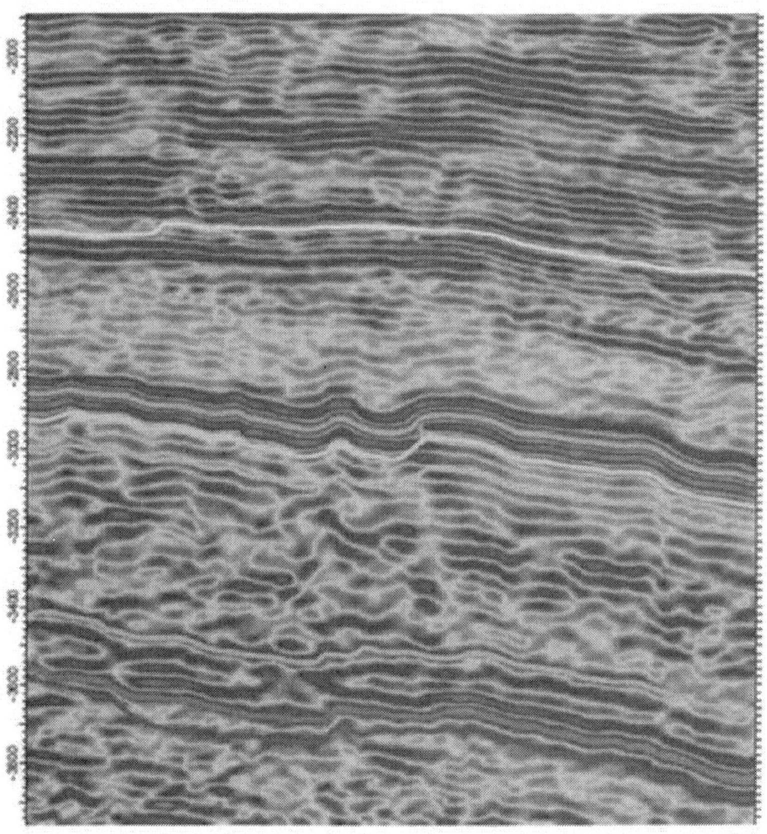

(Source: Erdwärme BayernGmbH & Co. KG)

**Figure 4.** Interpreted seismic reflection cross section with interpreted faults highlighted.

## 4.2. Gravity survey

Gravity survey in geothermal exploration defines the lateral density variation related to deep magmatic body, which may represent the heat source. These anomalies can be created by different degrees of differentiation of magma or variation in depth of crust–mantle interface which creates also depth variation of isotherms. This survey is simple and easy to be carried out using gravimeter. Once the survey is done, some processing parameters should be taken into consideration. More details about these parameters can be obtained from Seigel [9]. Figure (5) shows an example of gravity surveys in geothermal sites in Japan. It shows that geothermal areas always have low gravity anomalies due to heat sources which change the physical properties in the subsurface rocks.

http://dx.doi.org/10.5772/61679

On the other hand, gravity monitoring surveys in geothermal areas are used to define the change in groundwater level and for subsidence monitoring. Fluid extraction from the ground which is not rapidly replaced causes an increase of pore pressure and hence of density. This effect may arrive at surface and produce a subsidence, whose rate depends on the recharge rate of fluid in the extraction area and the rocks interested by compaction. Repeated gravity monitoring associated with weather monitoring may define the relationship between gravity and precipitation which produces the shallow groundwater level change. When gravity is corrected by this effect, gravity changes show how much of the water mass discharged to the atmosphere is replaced by the natural inflow. The underground hydrological monitoring done by gravity survey is an important indication of the fluid recharge in geothermal systems and the need of reinjection.

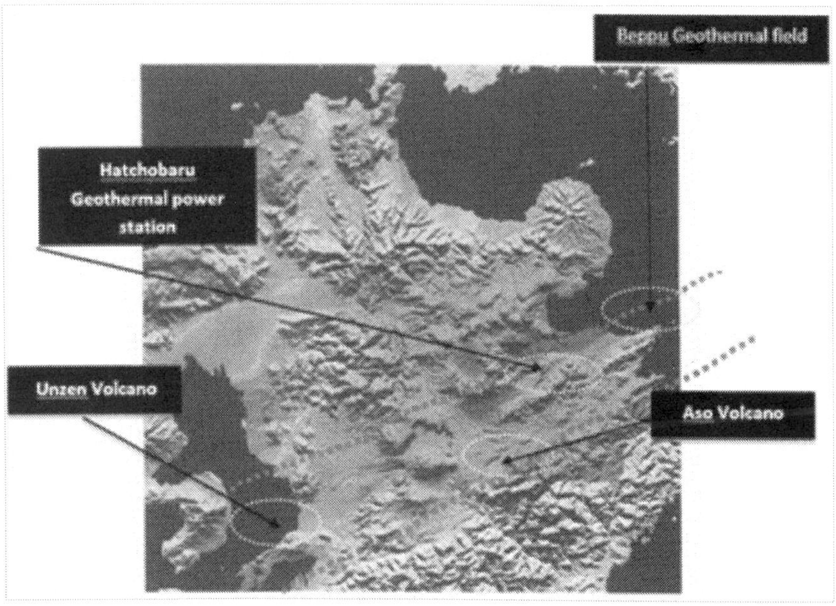


**Figure 5.** Gravity anomaly map of four geothermal sites, Japan [5].

### 4.3. Magnetic surveys

The earth has a primary magnetic field, which induces a magnetic response in minerals at and near the earth's surface. By detecting spatial changes of the magnetic field, the variations in distribution of magnetic minerals may be deduced and related to geologic structure. In geothermal exploration, each magnetic mineral has a Curie temperature, above which it loses its magnetic properties. This phenomenon is used to detect zones which are magnetically featureless, due to destruction of magnetite in near-surface rocks by hydrothermal alteration.

Figure (6) shows an example of Curie depth map of Sinai Peninsula and its relation to heat flow areas [1]. The map shows law Curie depth surface with high heat flow. The usefulness of magnetic surveys in geothermal exploration is controversial. On the other hand, magnetic method can be used to detect the subsurface structure within which the geothermal reservoir is build.

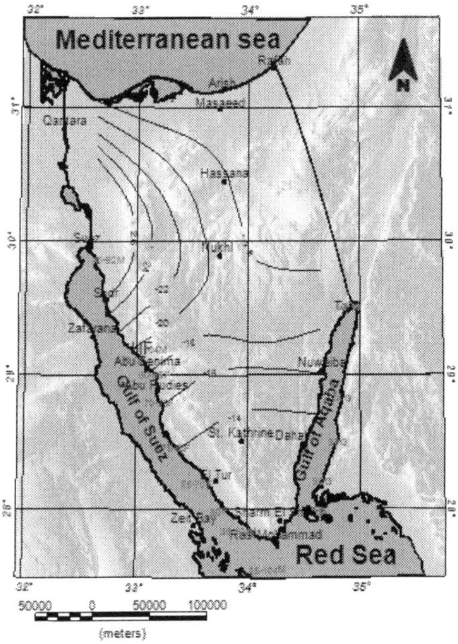


**Figure 6.** Curie depth contour map of Sinai Peninsula deduced from magnetic data. Background is topography relief. It can be recognized that, high heat flows match well with low Curie depths, indicating high geothermal potential [1].

### 4.4. Electrical/electromagnetic surveys

Most electrical/electromagnetic methods are used to measure the electrical resistivity of the subsurface rocks. Resistivity in the earth is often largely affected by electrical conduction within waters occupying the pore spaces in the rock. Consequently, resistivity varies considerably with porosity. Temperature and salinity of interstitial fluids tend to be higher in geothermal reservoirs than in the surrounding rocks. Consequently, the resistivity of geothermal reservoirs is generally relatively low.

In electrical methods, current is injected into the earth and the potential difference from which the subsurface resistivity can be obtained is measured. On the other hand, electromagnetic methods are a tool for determining the electrical resistivity distribution in the earth by means of surface measurements of transient electric and magnetic fields. These fields can be naturally

http://dx.doi.org/10.5772/61679

or artificially generated. These methods are more suitable for measuring the low resistivities of geothermal reservoirs than the above mentioned electrical resistivity methods. Furthermore, in geothermal areas, the surface resistivity is sometimes so high as to prevent current from entering the ground, and the electromagnetic methods, with a much deeper penetration, help eliminate the screening effect of very resistive surface rocks. Currents of varying frequency are transmitted into the ground, either via the electrodes as in the electrical methods, or by induction loops. Mobile stations measure, at several points, the electrical and magnetic fields created by this transmission. Comparison between these fields enables the resistivities of the underlying formations to be obtained, as a function of the frequency used that is as a function of the depth, as in the magnetotelluric (MT) soundings.

The magnetotelluric (MT) method responds to the earth's electrical resistivity structure [10]. The method involves taking a time series recording of natural, low-frequency, orthogonal electric and magnetic fields at the earth's surface, then interpreting the data in the frequency domain. Natural fluctuations in the earth's magnetic field are generated by lightning, ionospheric resonances, or variations in the solar wind. These fluctuations induce electric currents (or telluric currents) beneath the surface of the earth. The ratio of the electric field to the magnetic field in the induced electromagnetic (EM) wave is a function of the frequency of the signal and the bulk electrical resistivity of the ground. Lower-frequency magnetic fluctuations induce currents through a greater thickness of ground (Figure 7). Recording data over a wide frequency spectrum effectively gives information about a great thickness of ground. Lower-frequency records (i.e., information about greater depths) require longer collection times.

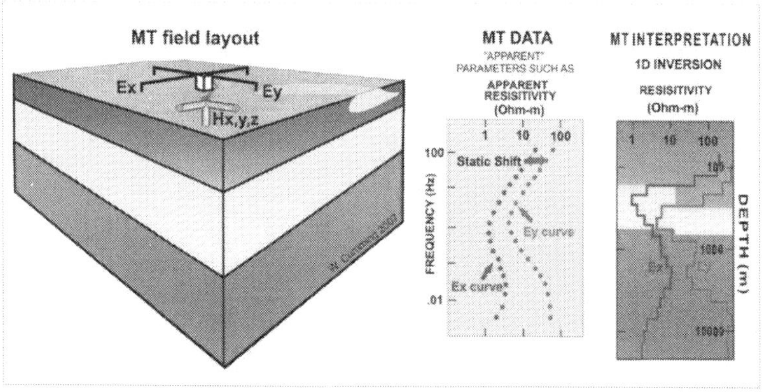


(Source: Harbour-Dom GmbH, Germany)

**Figure 7.** MT station layout and skin depths for natural electromagnetic waves depending on frequency, Low frequencies respond to deep structures, high frequencies respond to shallow structures.

The MT method is one of the very few geophysical techniques that can provide information about rock units deeper than about 1 km. This makes it useful for geothermal exploration, where target depths are typically in the range of 1–3 km for convection-dominated geothermal plays and even deeper for conduction-dominated plays. The MT method is particularly useful for convection-dominated plays because it can potentially imagine low resistivity and low permeability smectite clay units that often cap high-enthalpy geothermal reservoirs [6]. For this reason, the MT method is often used to reduce uncertainties about reservoir depth, geometry, and areal extent.

MT surveys can be performed at a regional scale. In these cases, the station spacing may be less than one per square kilometer. It is usually more cost effective to identify a prospective area with other methods and then conduct an MT survey with relatively high station spatial density in that area, with perhaps as many as 10–15 stations per square kilometer.

Magnetotelluric data are normally interpreted through an "inversion" process, whereby a semiautomated algorithm determines the simplest and most likely "apparent resistivity" structure consistent with the collected data. Inversions can also be carried out in 1D, 2D, or 3D, referring to both the spatial distribution of recording stations and the dimensions of the model simultaneously solved. A 1D inversion produces a vertical "sounding" from a single station; a 2D inversion, a profile from a line of stations; and a 3D inversion, a self-consistent block model from an array of stations [11]. Higher-dimensional inversions require significantly greater computing power and time to complete.

Inversions might be carried out by the MT contractor or by an independent third party. Inversion algorithms typically need to be constrained in some way, usually through limiting the allowable number of discrete layers and/or the depths between layers. For this reason, inversion results are subjective because they depend on input from the data processor. The results from 1D, 2D, and 3D inversions can differ significantly from each other for the same set of data, because the models depend on the dimensionality and complexity associated with the magnetotelluric responses. The resolution and accuracy of inversion models in terms of both depth and apparent resistivity decrease with depth.

The results of magnetotelluric inversion are normally presented as apparent resistivity on 1D soundings, 2D profiles (Figure 8), or 3D block (Figure 9).

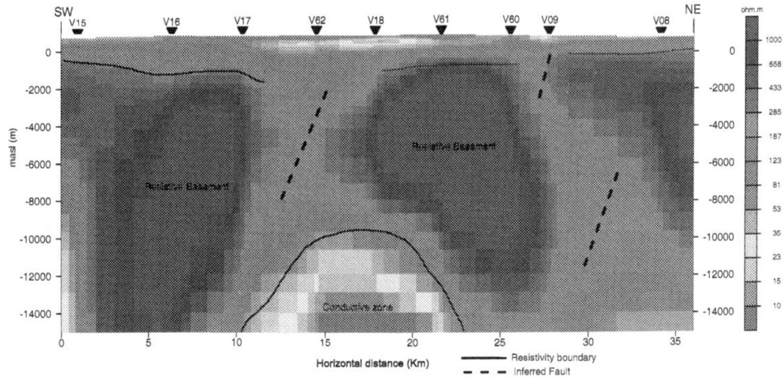


**Figure 8.** MT cross section profile showing the conductive zone at 13 km [8].

http://dx.doi.org/10.5772/61679

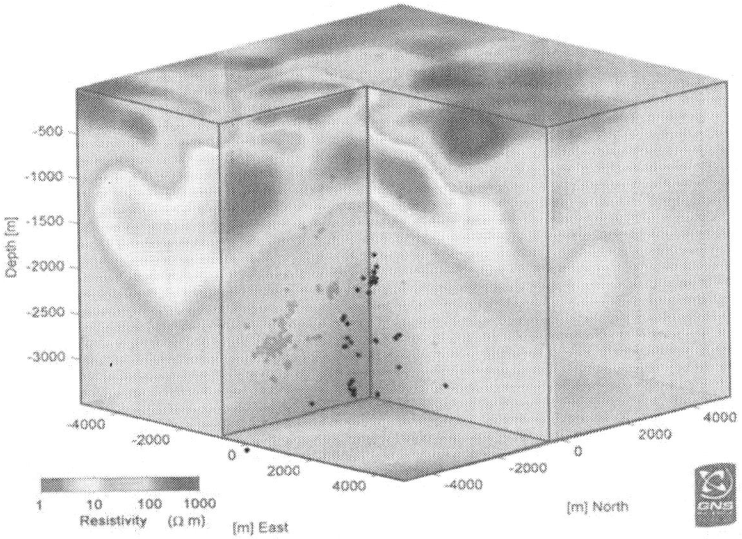


(Source: GNS Science, New Zealand)

**Figure 9.** MT resistivity block model.

## 4.5. Thermal measurement surveys

In geothermal research, the traditional geophysical methods mentioned above are used side by side with more specific techniques. Geothermal prospecting provides information on the thermal conditions of the subsurface, the aerial distribution of the earth's heat flow, and the location and intensity of thermal anomalies. To be more specific, geothermal prospecting allows us:

- To verify the existence of high-temperature fluids in areas without surface manifestations, but in which the subsurface structural and hydrogeological situation is favorable to hydrothermal circulation.
- To more precisely site deep drilling in areas that are considered potentially productive.
- To delineate the boundaries of geothermal fields that have been identified, to avoid drilling of dry holes in nonproductive areas.
- To acquire data for evaluation of the geothermal potential of the field.

Heat flow measurements are made by drilling small diameter (4 inches, 10 cm), shallow wells (< 300 m). Generally, heat flow is measured every 10–25 $km^2$. The geothermal gradient is obtained from temperatures measured with electric thermometers at various depths along a well. Temperature logging is quick and relatively inexpensive. The thermal conductivity of the rocks in the interval in which the gradient has been measured is usually determined by laboratory measurements on core samples. The product of the gradient and conductivity gives the heat flow.

## 5. Exploratory wells

The final stage of an exploration survey is exploratory well drilling. Usually, the final diameters of these wells are on the order of 8 inches (20 cm) or less, allowing the insertion of special logging tools to measure various parameters from the surface to total depth, and sometimes to carry out fluid production tests. A pump may be lowered into a shallow hot water well some hundreds of meters deep, and compressed air (gas lift) may be injected in deeper hot water wells. Since most geothermal reservoirs are made up of fluid-filled fractures, it is essential that an exploratory well intersects as many fractures as possible. Since natural fractures are related to tectonic activity (folding and faulting), the siting of exploratory wells is greatly dependent on our geologic interpretation of the local structural conditions.

## 6. Conceptual model

As indicated above, the preliminary survey, exploration, and test drilling phases of a project are all about defining, refining, and testing a "conceptual model" of the geothermal system under investigation; a conceptual model is the schematic representation. A good conceptual model should encapsulate the geological framework, heat source, heat and fluid migration pathways, reservoir characteristics, and surface geothermal features, and should be consistent with all available data and information. The conceptual model is continually refined as each new set of data is collected and assessed, with each refinement adding a new level of detail or confidence to the overall model.

An initial conceptual model should be developed at the earliest stages of the geothermal project. At this time, the model will necessarily be quite crude, perhaps illustrating little more than a generic representation of the expected geothermal play type. The model should then be regularly updated as new data become available to ensure the model respects and remains consistent with all known information. In this way, the most current conceptual model should incorporate all available exploration data. By the end of the exploration phase, the conceptual model should be of sufficient detail to allow an estimate of reservoir depth, temperature, and geometry with sufficient confidence to justify and site wells for the test drilling phase.

The conceptual model can be illustrated with maps, 2D cross sections, or 3D block models. These might be simple free-form drawings at the early stages of a project, but will develop into robust geological models as more information is incorporated. Cross sections should be created at the same scale as the maps that underpin them, preferably with a 1:1 ratio between horizontal and vertical scales. All diagrams should include a representation of the assumed heat source, an estimate of the subsurface temperature distribution (isotherms), some indication of fluid flow directions, and a representation of the expected geothermal reservoir, even if these are only approximate.

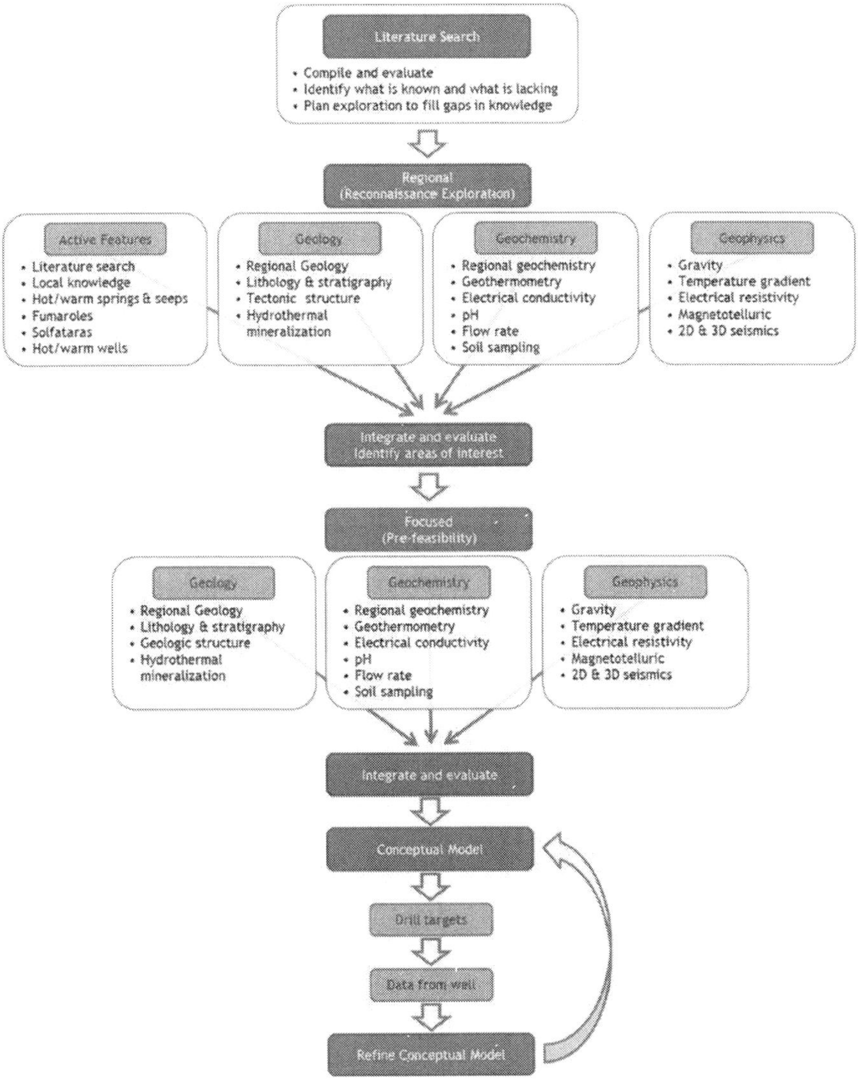


(Source: GeothermEx Inc., California)

**Figure 10.** Flowchart showing project stages with typical data acquired and integrated into the conceptual model.

A good conceptual model provides clear evidence that the explorer has considered and integrated all available data. Nothing in the conceptual model should contradict the data presented elsewhere, unless a clear rationale is provided. The conceptual model demonstrates a justifiable understanding of the geology, temperature, and fluid pathways within the geothermal system. By utilizing the conceptual model, the explorer can select sites for the test drilling phase that maximize the chances for a successful well based on all current data.

All exploration data should be integrated into a conceptual model of the geothermal system under investigation. This model must respect and be consistent with all known information. Figure (10) provides a flowchart of typical data that may be used to build and develop the model. The model needs to be of sufficient detail to allow a first-pass estimate of resource temperature and size and, in the test drilling phase, is used to target deep, full-diameter wells toward particular lithological units and/or structures that are judged most likely to deliver commercial rates of geothermal fluid at commercially viable temperatures.

## Author details

Essam Aboud[1,2*]

Address all correspondence to: eaboud@gmail.com

1 Geohazards Research Center (GHRC), King Abdulaziz University, Jeddah, Saudi Arabia

2 National Research Institute of Astronomy and Geophysics (NRIAG), Cairo, Egypt

## References

[1] Aboud E, Salem A, Mekkawi M (2011) Curie depth map for Sinai Peninsula, Egypt deduced from the analysis of magnetic data. Tectonophysics 506:46–54

[2] Ford TD, Pedley HM (1996) A review of tufa and travertine deposits of the world. *Earth Sci. Rev.*, 41:117–175

[3] Giggenbach WF (1975) A simple method for the collection and analysis of volcanic gas samples. *Bull. Volcanol.* 36:132–145

[4] Hochstein MP (1990) Classification and assessment of geothermal resources. In: Dickson, M.H. and Fanelli, M., *eds.*, *Small Geothermal Resources: A Guide to Development and Utilization*, UNITAR, New York, pp. 31—57

[5] Koichi Tagomori, Enjang Mustopa, Hisashi Jotaki, Hideki Mizunaga and Keisuke Ushijima (2005) Proceedings, Thirtieth Workshop on Geothermal Reservoir Engi-

neering Stanford University, Stanford, California, January 31-February 2, 2005 SGP-TR-176

[6] Melosh G, Cumming W, Casteel J, Niggemann K, Fairbank B (2010) Seismic Reflection Data and Conceptual Models for Geothermal Development in Nevada.

[7] Pentecost A (1995) Geochemistry of carbon dioxide in six travertine-depositing waters of Italy. Journal of Hydrology 167: 263–278 http://dx.doi.org/10.1016/0022-1694(94)02596-4

[8] Peter W, Aboud E, Abdelwahed M, Cherrington J, Hoeberechts J, Catherine K, Lindsay J, Moufti M (2013) Magnetotelluric imaging of the Northern Harat Rahat volcanic field, SCIENTIFIC MEETING NOVEMBER, Jeddah, KSA 17–18

[9] Seigel HO (1995) A guide to high precision land gravimeter surveys, Scintrex Limited, 222 Snidercroft Road, Concord, Ontario, L4K 1B5

[10] Simpson F, Bahr K (2005) *Practical magnetotellurics.* Cambridge University Press, Cambridge, 270 pp

[11] Siripunvaraporn W, Egbert G, Lenbury Y, Uyeshima M (2005a) Three-dimensional Magnetotelluric inversion: data-space method. Phys Earth Plan Int 150:3–14

[12] Wright PM, Ward SH, Ross HP, West RC (1985) State-of-the-art geophysical exploration for geothermal resources. Geophysics 50:2606–2699